The Undersea
PREDATORS

The Undersea
PREDATORS

by **Carl Roessler**

Facts On File Publications
New York, New York • Bicester, England

To Jessica —
Sharing our life is my inspiration

All photographs are by the author.

This book was produced in conjunction with
PBC International. Inc.

Library of Congress Cataloging in Publication Data

Roessler, Carl, 1933-
  The undersea predators.

  Includes index.
  1. Marine fauna—Food.   2. Marine ecology.
3. Predatory animals. I. Title.
QL121.R64   1984   574.5′3′09162   84-1547
ISBN 0-87196-893-2

Library of Congress Catalog Card No.: 84-1547

ISBN: 0-87196-893-2

Printed in Hong Kong
by Toppan Printing Co. (H.K.) Ltd., Hong Kong
10 9 8 7 6 5 4 3 2 1

Table of Contents

Foreword

I first began to appreciate Carl Roessler's writing skill when I read his splendid book, *Underwater Wilderness*. He combines scientific veracity with poetic, passionate imagery. His years of observation have honed his descriptive skills. He has wandered coral reefs for so long that the welter of shapes and colors, the rushing movements of fishes, the momentary glimpses of action, no longer confuse him. His position as owner of a scuba-diving travel service has afforded him the luxury of diving all over the world, with enough time underwater to concentrate on trying to answer questions puzzling all divers when underwater. He describes situations when he spent perhaps an hour following an octopus, minutely chronicling its color and behavior changes as first it feared him, then became so used to his presence that it literally posed for his camera.

His experiences underwater are legion. It seems to be part of his job to lure sharks to the boat so that his diver clients can photograph them. He describes an "occupational hazard" involved in cleaning up after a shark-filming session: The process of baiting the sharks, the bubbles and sounds of excitement as the divers took their photographs — all this attracted a surprise guest, a 15-foot hammerhead shark. Fortunately it settled for some leftover fish-heads which inadvertently dropped from Roessler's hands as he stared, stunned, at the huge animal just five feet from him.

Throughout the book Roessler tries to evoke scenes uncolored by human preconceptions. He constantly admonishes the reader (and diver) to shed his human emotions of pity, rage, and fear, and to enter the sea as naturally as the fishes. Yet his images of being alone at night with a failed underwater light, surrounded by the inky darkness of the sea, cannot help but bring chills to the spine.

Roessler's work is characterized by a scholarly accuracy of detail, yet he does not claim to be a scientist. This permits him to conjecture about the "motives" and behaviors of undersea animals without the limitations of science — the need to watch every word for fear of controversy. Roessler is free to soar above mere observation — he wonders, conjectures, tries to put himself into the skin of the predator as it stalks its prey. Why do fishes school? Why are predators such as jacks and barracuda more successful at dawn and dusk than during the rest of the day? Sometimes one of the sea's mysteries seems beyond conjecture. Why should fishes change their color at night even though it is so dark that this color change cannot be seen?

But the most important virtue of Roessler's work is not the poetry of his language or the beautiful images he evokes. It is his vast experience underwater. He has actually seen the kinds of underwater marvels that most of us only read about. He infuses into his descriptions those reactions and personal vignettes that recreate his momentary confrontations with morays, squids, and other undersea inhabitants. He describes the *sounds* made by parrotfishes as they bite off chunks of coral, and the darkening of the water as a school of these fishes defecate the now thoroughly chewed coral as clouds of fine sand.

Carl Roessler describes the undersea world with compassion. The reader cannot help but appreciate the sea's complexity and awesome beauty. Perhaps some of this new understanding and appreciation will help us to gain the perspective we need to manage the sea's resources before they are irreparably destroyed.

EUGENE H. KAPLAN Director
Hofstra University
Marine Laboratory

Introduction

Whenever the words "undersea predators" are spoken we have long been conditioned to conjure certain images — principally, that of gray and sinister sharks prowling both the depths of the sea and the margins of our racial memory. These sinister images have been with us since man for the first time beheld the sea's prototypical predator. Dark, lethal, and efficient, the shark is simply the sea's most perfect symbol of the never-ending search of predator for prey.

Yet in the complex web of marine society the shark is but one of many creatures whose lives are entwined with each other in that intricate ballet of life or death, kill or be killed, eat or be eaten — the hunt. When we see, for example, a shallow coral garden gleaming under a tropical sun, it strikes the same gentle chord of perception for us as does a terrestrial flower garden. But that perception could not be more wrong: all of those plantlike structures that compose the marine "garden" are actually animals. Moreover, they are all extremely efficient and well-adapted predators who have evolved brilliant solutions to their particular environment's supply of food and shelter.

Ours is the first generation of man to enter the undersea world in numbers — thousands, even millions of us observing the creatures of the sea not in the laboratory but in their natural habitat. We are privileged to be the pioneers on this final earthly frontier. In the beginning, only scientists and adventurers entered the kingdom of the sea, armed mainly with curiosity and courage, often using the most primitive and dangerous of equipment. Today, groups of vacationers with snorkels or modern scuba gear and cameras are discovering the world's coral reefs, finding that the society thus revealed is at once exceedingly complex and beautiful beyond anything in our terrestrial experience. The webs of misconception, prejudice, and ancient terror are being cleared, understanding replacing assumption and taboo. Yet there is so much of the sea's tapestry we have yet to see.

The subject of predation and its role as a force in the marine environment is central to all we have learned. The almost-incomprehensible variety of hunting techniques which marine creatures have evolved gives us insight into the daily round of life on the reef. As we learn more, the reef-dwellers take on an unexpected individuality worthy of citizens in an intricately balanced social and biological system. This volume examines a kaleidoscope of unsuspected predators, large and small, revealing the amazingly different yet invariably effective techniques they have evolved in their struggle for survival. Predation in a real sense is the engine that drives the continuous spectacle of life on the reef. It is the primary reason animals move, take shelter, develop specific habitats, and seek or avoid each other.

Some will say, "Haven't you forgotten the activity of perpetuating the species? Isn't breeding new life even more fundamental?" My answer would be that breeding activities generally occur during certain seasons or under certain circumstances, and are in some species limited to a single day or a few days out of an individual's entire life cycle.

Predation, on the other hand, is a constant underlying all of the reef's activities. As we shall see, even during breeding activities the hunt goes on; indeed, the breeding activities themselves have evolved with strict attention to the constant danger of predation.

The specific method used by any species in its pursuit or avoidance of predation helps us place it in the overall scheme of reef activity; a spectrum of otherwise very different animals will share a basic approach to the capture of their food. Since all marine animals pursue one or another method of predation, our subject matter is as broad and as varied as the sea itself.

For our purpose, certain broad categories of predation offer a natural framework for organizing our survey. Some creatures are rooted firmly in one spot throughout their entire lives; their methods of gathering food contrast vividly with those of other creatures that rely on blinding speed, or silent ambush, or camouflage to capture their prey.

In our journey we'll meet some unusual creatures, whose evolution has provided them with surprising variations on common methods of hunting. As we meet this spectrum of predators, we'll discover a natural hierarchy that exists among them. In the end, perhaps to our sorrow, we'll find that the most feared predator of all is man.

This is, then, a journey of imagination among the citizens of the world's great reefs. If you can, imagine yourself not as an observer, there for an hour and gone, but as a reef-dweller occupying your own niche, for a lifetime, in undersea society.

Seeing the world from that perspective may give you a new insight into the future of our planet.

Hunters Who Cannot Move

The first family of predators we shall consider are those which most observers never suspect of predation. Across the sunlit shallows of the tropical world are what seem to us gardens of coral. As we look at these gardens our perception of their life is unavoidably influenced by our experience on land. To the mind of terrestrial man, a bed of corals has as its closest analogy a flower garden. Both are beautiful, both are characterized by a profusion of color and form, both are "rooted" in the ground. Naturally, then, a bed of brilliantly colored corals just beneath the surface of the water evokes the flower-garden image, and many of us who write about the sea's creatures further this impression by referring to these communities of coral as "coral gardens."

The truth about corals and other sessile (rooted) sea creatures is far more intricate and complex.

To begin with, we must recall that the sea is like a soup. Even at its clearest, seawater contains quantities of phytoplankton and zooplankton. Phytoplankton are those complex communities of single- and multi-celled, free-floating creatures which collect sunlight and obtain their nourishment via photosynthesis. These plantlike cells are a principal source of our planet's oxygen supply. Most zooplankton, on the other hand, are often tiny creatures who ingest phytoplankton and other zooplankton for their food.

In order for such sessile creatures as corals, *Tridacna* clams, sponges, and tunicates (sea squirts) to have access to this plankton soup as a reliable food supply, the soup must be stirred. The motion of the earth provides this activity through wind and water currents which move the plankton in endless cycles about the sea's great basins. Plankton and ocean currents are the key to survival for the stationary creatures of the coral reef.

Consider the basic body plan of a coral polyp. The body is a soft, radially symmetrical cylinder containing a stomach and digestive tissues, plus reproductive tissues. Atop this cylinder is mounted a central mouth. Around the mouth on the upper surface (or at least around the outer edge) of the cylinder is a ring of food-gathering tentacles.

There may be eight or a multiple of eight tentacles (octocorals) or six or a multiple of six tentacles (hexacorals). In some corals these tentacles are smooth and snaky, while the tentacles of others have tiny fingers branching from them at right angles. The tentacles of corals and their relatives such as anemones are armed with chemically activated, springloaded stinging cells, called cnidocytes, which are tipped with neurotoxins. While the poison darts, called nematocysts, in these stinging cells usually cannot penetrate human skin (if they do, the pain is temporary), they are highly lethal to marine organisms such as zooplankton or, in the case of anemones, to fishes. The tentacles capture food that is carried along in the current-driven waters and pass the food to the central mouth to be ingested by the coral.

Many corals also have a complete back-up system to survive even a complete absence of current-borne food. Embedded in the tissue of coral polyps are microscopic algae cells called zooxanthellae. These tiny plant cells gather sunlight and the coral's waste, and photosynthesize food and oxygen. Thus, the coral and zooxanthellae nourish each other even without other food. It is, in fact, the zooxanthellae which usually give color to the coral; coral polyps themselves are colorless.

Another profoundly important chemical activity of the coral polyp is the secretion of limestone (calcium carbonate) in the form of a protective cup about its body. When a coral is not extending its tentacles to feed, its body is withdrawn into and protected by the stony cup.

The cup each coral polyp secretes is extended by the polyp over the years. The cumulative lengthening of thousands of cups by succeeding generations of corals creates a coral head. The base of any stony-coral structure is composed of ancestral skeletons, with the current generation of live polyps inhabiting their cups on the surface of the structure. When you think of the massive size of many coral heads, you begin to appreciate both the numbers and the crucial building role of corals on a reef. Vast accumulations of coral heads form the world's coral reefs. Whole tropical islands are composed of immense coral reefs rudely thrust above the water's surface by the forces of plate tectonics or glaciation.

You soon begin to appreciate the astronomical numbers of coral polyps involved in these dry recitations of fact. Spend a moment on a snorkel or scuba dive and try to count the coral polyps on a single coral head. There will be at least hundreds and usually thousands of individuals. Each of those individuals represents the latest generation in a history of builders that may also number in the hundreds or thousands. You can easily imagine a million polyp lives expended just in the formation of one coral

structure smaller than your hand.

Now back off a bit. Look at several coral heads, then hundreds and thousands covering the span of reef you see.

Then imagine countless coral heads lost over eons in the vast rock bulk beneath the reef you are watching.

The number of polyps is simply staggering. It would take the loquacious scientific genius of an Isaac Asimov or a Carl Sagan to evoke them properly. I simply stand in awe.

Corals reproduce both sexually and asexually. Each method performs a valuable function. Asexual reproduction (or budding) occurs when an individual polyp develops a bud which grows right next to it. This is invaluable to the dome or reef-building corals. If you were to draw on a balloon a number of coral polyps, then inflate the balloon, you immediately see that the role of budding is to fill in the gaps as the dome grows larger.

Sexual reproduction releases eggs and sperm into the open water. The resulting pelagic larvae enable corals to colonize new regions far downstream, or to repopulate areas of natural destruction.

Interestingly, in certain areas where men have dynamited coral reefs for fish or construction, new corals seem to have difficulty establishing themselves in the moon-barren craters.

Among the 2,500 species of coral there is an astonishing variety of shapes, sizes, and colors in the structures they form.

There are vast castles of tan *Montastrea cavernosa* corals; tortured edifices of fire coral (millepora); dainty pink or purple *Acropora* with their delicate antler-like branches; and the soaring stonehenges formed by huge beds of elkhorn coral.

Then there are the gorgonians. Those which live in the shallows, where the wave-surge is their frequent companion, have a flexible skeleton of wiry gorgonin which can withstand countless wave-surges without breaking. Other gorgonian corals are limited to rocky canyons in vast reef structures, canyons through which tidal currents flow. In the South Pacific, this environment can produce fans of brilliant yellow, several feet across and always set at right angles to the current flow. A third type hides its brilliant reds, its intense purples and pinks, in the sunless depths below 150 feet.

The royalty of coral are the immense soft corals of the Indo-Pacific. Some of these colonies reach six feet in height, and are found in a rainbow of subtle, lush colors such as pink, purple, or soft gold. Like their stony-coral cousins, the colony is composed of thousands of individuals. Rather than limestone cups, these soft coral polyps form huge, inflatable bags to elevate themselves into the food stream. When the currents stop, or when the sun is high and the zooxanthellae active, there is no need for the colony to inflate its bag structure with water. Some colonies look quite pathetic when deflated.

As much as they are builders of the reef community, corals are also prey. As we examine our families of predators in pages to come, we'll find that many of them prey on the coral polyps in a variety of ways. Together with plankton, coral constitutes the broad base of the food chain.

First and foremost, however, these simple but still amazingly complex creatures are efficient predators whose morphology and techniques have withstood the onslaught of countless centuries.

Relatives of the true corals are the anemones. Anemones are *Anthozoans*, sometimes called "flower corals." Anemones have very large bodies (sometimes two to three feet across), hundreds of tentacles, and an appetite for more complex foodstuffs than plankton — fish, for example. The anemone's toxic nematocysts quickly paralyze any fish unlucky enough to brush a tentacle, and I have seen a foot-wide anemone devour a foot-long fish. On that occasion the fish's body was bleached white where the lethal tentacles had lashed it. So powerful was the impact of the nemotocysts' toxin that the fish shuddered briefly and moved no more. . . .

Cerianthid anemones are unusual in that they hide their delicate bodies in tubes constructed of nematocysts, mucus, and sand; they can also completely withdraw beneath the sand of the sea floor. Most cerianthids are nocturnal, and can be spotted by their characteristic tube and their long, slender, filament-like tentacles.

Another family of sessile animals are the frequently very lovely segmented polychaete worms. One variety, the sabellid worms, hide their bodies like cerianthid anemones in tubes of sand and mucus; others, the serpulid worms, hide in secreted limestone. Their appendages for food-gathering and respiration, however, are not only visible but often quite beautiful. The well-known turkey worms of the Caribbean are the most commonly observed sabellids, while the "Christmas tree" serpulids are found on coral reefs throughout the tropical world. These worms spread delicate, feathery arms to which plankton, larvae, and debris adhere. Ciliary action moves the captured particles to the central mouth.

Several unusual polychaetes, including *Loimia medusa* and *Eupolymnia nebulosa*, have long, slender filaments which resemble fishing lines. Divers who spot these white lines are nonplussed to see them withdraw at a touch. The lines are covered with a sticky mucus. When the lines are drawn in, the worm feasts on the bits of food that have adhered to the mucus.

Moving to a totally different group of sessile predators, we discover another highly successful predatory technique. This method is known as filter feeding, and it is practiced in varying forms by sponges, *Tridacna* clams, and the ubiquitous tunicates (sea squirts), salps, and barnacles.

Barnacles "pulse" regularly, popping out of their protective shells for just an instant to filter the water to strain out food, then disappearing back inside. These pulses occur on the order of every second, which makes the barnacle a very active predator indeed.

Sponges are most abundant in the coral reefs of the Caribbean. On Pacific coral reefs they are far less prevalent, perhaps crowded out by more (or more aggressive) species.

The Caribbean is the home of the giant barrel sponge and a host of smaller, brightly colored varieties. The Cayman Islands seem to have the greatest concentration of sponges, a veritable arsenal of gunbarrel-like tubes in crimson, gold, blue, and green, and larger "cannons" occurring in burgundy or roan. Some barrel sponges are so massive that two divers can squat down and disappear within a single specimen.

Scientists have long debated whether sponges are single animals or colonies. Without question, a sponge is an aggregation of specialized cells, often stiffened by spicules of calcium carbonate embedded in the mass. Certain flagellated cells flutter ceaselessly, creating a current which is expelled out the mouth of the tube. If, for example, you release a bit of yellow fluorescein dye near the outer surface of a tube sponge, the sponge will soon expel billowing clouds of the yellow dye out the end of its tube. Had any food particles been present near the cilia, they would have been captured by the maze of cells in the sponge's body wall.

Encrusting sponges such as *Crambe crambe* form a gaily colored skin-surface on any solid substrate, such as dead coral. On these specimens you can see patterns of internal tunnels leading to the various oscules.

On several occasions I have witnessed the phenomenon of the "smoking sponges." In the Caribbean this usually occurs in late August. The "smoke" is the simultaneous release of countless sperm and eggs from the sponges. By releasing the sperm and eggs at a given time, the sponges maximize the chances of fertilization and hence the continuation of the species. The larger sponges will spew copious clouds of white, smoky genetic material, which drastically reduces visibility in the surrounding waters.

Other spectacular filter-feeders are the *Tridacna* clams, which can reach a length in excess of five feet and a weight in excess of half a ton. Smaller *Tridacnas* with brilliantly colored mantles are found throughout the South Pacific, but many of the giant clams have fallen prey to fishermen. *Tridacna* meat is a great delicacy in Asia, and Taiwanese and Korean fishing boats have harvested giant clams from Australia to the Maldives. Without enforced regulation, these spectacular giants will disappear, another tragic monument to man's greed and thoughtlessness. To a diver, a world without these gentle behemoths is like a world without elephants. Yet both of these classic giants may not survive our century.

Another successful family of filter feeders are the tunicates, or sea squirts. Tunicates are easily recognized by their pairs of inlet and outlet openings, though otherwise their bodies are found in a variety of shapes from ball to kidney bean.

Some tunicates are elevated on stalks to perch above the surrounding coral competitors. Usually larger stalkless varieties attach to coral outcroppings to elevate themselves into the food stream. These latter are kidney-bean-shaped, with one opening on the end of the body and another atop the body near its center. Still other tunicates, tiny beauties of orange or golden yellow, gather in clusters of 20 to 50 individuals like bouquets of predatory flowers. These smaller tunicates are flattened spheroids, with one central opening and a second just off-center adjacent to the first.

Rarest of all are the salps, pelagic relatives of the sessile tunicates which have forsaken a "rooted" existence and formed huge colonies which float freely in the open sea. Thousands of individuals, *Pyrosoma*, will form long, hollow pink cylinders that have been known to reach 60 feet in length and 8 feet in diameter. These colonies greatly resemble the popular toy known as a "Slinky," and their fluid, coiled motion is eerily reminiscent of that toy. Rooted together in a pink spaceship, floating endlessly in the open sea, these are the Flying Dutchmen of the tunicate world.

All of these rooted creatures have evolved highly successful techniques of predation despite the fact that they can neither swim nor crawl. Forever fastened to their place in the reef world, they prosper by sampling the current-driven food stream. From the plankton's point of view they are far less a garden than a mortally dangerous obstacle course, lined at every turn with evolution-proven predators.

The azure vase sponge has an ethereal iridescence as it feeds upon the passing food stream (Belize, Caribbean).

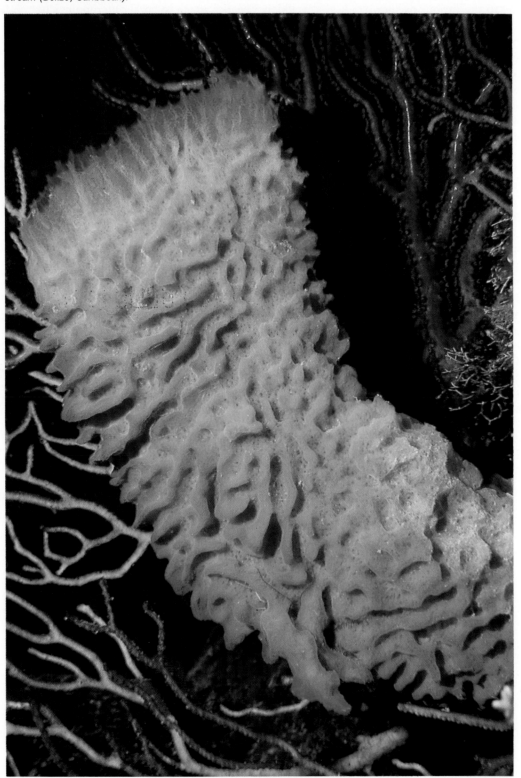

In cold water, strawberry
corals carpet stony walls with
tapestries of soft color
(Galápagos).

The pink hydrocoral Stylaster *builds colonies of lacy elegance (Coral Sea, Australia).*

Some coral heads achieve enormous size, commanding an entire section of coral reef (Ponape, Micronesia).

An unusual cluster of tube worms spread their miniscule, feathery plumes to capture plankton (Philippines).

An unfortunate grouper is lashed by the lethal tentacles of an anemone (Coral Sea, Australia).

The tunicate (sea squirt) resembles a U-shaped tube with a strainer (Philippines).

A brittle starfish hides its vulnerable body within that of a colorful tube sponge, reaching only its delicate arms to feed (Caribbean).

Colonies of soft coral polyps share a central sac which inflates with water, elevating the polyps upward to reach passing food (Fiji).

In the Coral Sea the serpulid worm Spirobranchus giganteus *is found in a golden yellow color (Australia).*

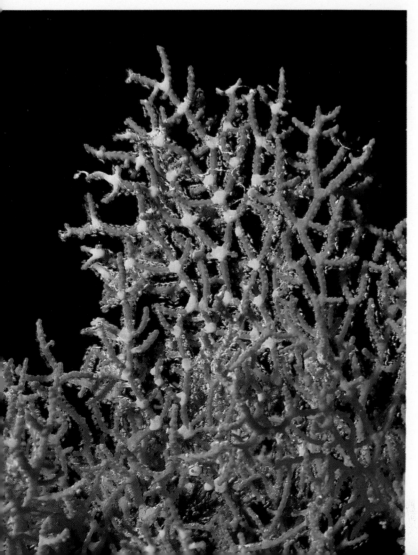

These thickets of tiny gorgonian polyps fill shadowed crevices in the reef mass (Maldives Islands).

19

Tunicate colonies take a variety of forms, such as this artistic design on a dark wall (New Zealand).

Here the coral polyps can clearly be seen in their individual limestone cups (Australia).

These coral polyps are found
on coral walls swept by
strong currents, offering a
rich food supply (Philippines).

The delicate Cerianthid
anemone withdraws entirely
beneath the sandy sea floor if
approached (Philippines). ▶

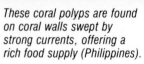

Soft corals occur in the Indo-
Pacific, but are not found in
the Caribbean (Philippines).

2

Those Who Move, But Slowly

Moving slowly among the sessile predators are a community of creatures that are seldom actually seen in motion. In many cases, their locomotion mechanisms are marginal, used principally to enable these predators to crawl from one feeding area to another, or from their daytime feeding position to their nocturnal shelter.

In chapter 1 we found that the sessile predator population was dominated by members of the large and variegated coral family. Similarly, we find that the slow movers are dominated by variously evolved members of the sea star family.

Most of us are familiar with the five-pointed star shape of most sea stars. What we may not realize is the remarkable number of variations on that theme which occur in the undersea world. Not only are there wildly different body shapes; some starfish, such as the notorious crown of thorns (*Acanthaster planci*), are such efficient predators that their proliferation is feared among marine biologists throughout the world.

Let's begin, though, with the common garden-variety five-pointed starfish. These vary in diameter from one inch to huge specimens of eighteen inches to two feet. Smaller starfish are found throughout the seas of the world, while the large specimens are found in two principal (and totally different) environments — the cold waters around the Galápagos Islands, and the tepid to warm shallow lagoons of the South Pacific atolls.

Some species devour molluscs by wrapping their arms about the bivalves and pulling the shells apart. While the mollusc's adductor muscles may hold for a long time, in the end the starfish's exhausts the mollusc; the shell wearily opens, and the starfish devours its prey.

Other bottom-dwelling starfishes use their tube feet to sift the coarse sand of the bottom and separate out the tiny molluscs which live in profusion on any sandy sea floor. Within this family of bottom-feeding starfishes we find several body-shape variants. The so-called bat starfish departs from the usual pattern of small central body and long arms. In the bat starfish, the body and arms form a flattened pentagram which maintains the five-pointed symmetry but makes the "arms" merely points on a solid body.

The singular pillow starfish looks like a bat starfish with a serious case of the bloats. In a pillow star, the body thickness may be one-half or more of its width. Many times, I've had divers approach me holding these vaguely star-shaped basketballs, obviously wondering what they've discovered. I turn the creature upside down, and they can see the star-pattern of tentacle-lined slots which define the animal's genealogy.

Other fascinating varieties of the starfish family are the many-armed stars. Of these, two come most quickly to mind — *Heliaster,* the many-armed starfish which is prolific in the Galápagos, and *Acanthaster,* destroyer of reefs.

The basic body color of *Heliaster* is black, but it is speckled with gaily colored spots. In different individuals these spots can be yellow, orange, red, or green; all of the spots on any one individual, however, are of the same color.

Where *Heliaster* is lovely to look at, *Acanthaster* is literally fearsome. Throughout history, regular population explosions of this immense predator have devastated vast areas of reef across the tropical Pacific.

Acanthaster is a many-armed starfish, with a dozen or more arms. Some individuals can be two feet across and weigh ten pounds when taken from the water.

These behemoths suffer few predators, since they are covered with thick, sharp spines up to two inches in length. The spines are armed with a toxin; once I was stabbed in a finger joint and suffered periodic pain and stiffness for nearly a year. While one would think it would require some large, savage predator to control the predation of this awesomely armed starfish, the two predators who do attack *Acanthaster* are, unexpectedly, otherwise harmless. One of these predators is a small shrimp; the other is the triton trumpet *Charonia tritonis,* a medium-sized mollusc with a highly prized shell.

During the most recent population explosion of *Acanthaster,* some observers blamed shell collectors for decimating the triton trumpet population, thus permitting the *Acanthaster* population to increase to terrible levels. Subsequent research, however, indicated that similar population outbreaks had occurred on average every 125 years. This was determined by analyzing the composition of cores extracted from the Great Barrier Reef. In these cores, layers rich in limestone spikes from *Acanthaster* occurred regularly. Is *Acanthaster* nature's way of pruning the reef to stimulate new growth? Or is there a more subtle purpose to this assault? Further research may provide answers, but at present our knowledge base spans a mere 15 to 20 years. You can be sure that this predator will be closely studied for decades to come!

The crown-of-thorns, unlike other starfish, eats coral polyps. It does so by flooding them with digestive juices, then devouring the softened polyps until only a bleached, white limestone skeleton remains. Having witnessed several reef areas devastated by these huge crown-of-thorns starfish I can attest to the moonscape they leave behind. One of my favorite reefs was literally stripped of all living coral shallower than 50 feet. During one infestation, my wife Jessica and I removed over 250 starfish from a reef line perhaps fifty yards long.

One of the great strengths of all starfishes is their ability to regenerate lost arms. Indeed, a severed starfish arm may itself regenerate an entire new starfish. Thus, early divers who cut up *Acanthaster* with knives to protect their reefs merely added to the plague. Only injection with formalin, or better, complete removal and burial ashore can break the skirmish line. We have done this successfully to protect particular reefs for divers.

Other members of the slow-moving starfish clan are far more elegant and graceful to our human eye than those mentioned thus far. The crinoids are a masterpiece of design. So successful is their morphology that it has remained essentially unchanged for 400 million years. Paleontologists find fossil crinoids from the Cretaceous period that are strikingly similar to those of modern reefs, except that the fossils were attached to the bottom. The long, graceful feeding arms are lined with numerous projections called pinnules, and the pinnules are lined with tiny tube feet. The crinoid uses its small array of walking legs, called the cirri, to crawl up upon a coral or gorgonian, where it spreads its feeding arms across the food stream carried by ocean currents.

I have witnessed demonstrations in which fluorescein dye was released just upstream from the arms of a crinoid, *Comanthina schlegeli.* The experiment showed that the complex arms of the crinoid act as a baffle, trapping the water amid the arms so the tube feet can capture tiny planktonic prey. The prey is then passed along the pinnule to a food groove in the long arm. Tiny cilia move the food down the groove for a distance of six to ten inches to a precisely matching groove in the main body. Cilia in this body groove carry the food to the waiting central mouth. This is a masterful, precise design which is one of the wonders of the reef. It is also highly efficient. Only in a mild to strong current are the arms of these crinoids fully deployed for feeding. When the current (and hence the food-stream) has ceased, the same individuals may be observed with their feeding arms gracefully folded inward.

While many species of crinoid remain in the same feeding perch for days or months, others are nocturnal. These creatures of the night remain deeply sheltered by day, moving up upon a feeding spot only after dark. Interestingly, many nocturnal species display bold barber-pole stripes on their feeding arms.

For those divers among you who have witnessed crinoids in a rainbow of colors on Pacific reefs in Australia's Coral Sea and the Philippines, color does *not* identify species. One species of *Comanthina* may occur in twenty or more effulgent colors from canary yellow to green to fire-engine red to powder blue.

An unusual, even eerie, starfish is the basket starfish, *Astroboa nuda*. By night, these starfishes climb atop corals and spread an astonishing complex of intricate arms in a great, circular, basketlike fan.

Divers who have done night dives know that in many tropical areas their lights attract copious quantities of small worm-like wrigglers that are a nuisance to photographers because they show up as blurs in their pictures. In these conditions, moving in to photograph *Astroboa* provides a frenetic display of predation. The wrigglers collide with *Astroboa's* complex of feeding arms by dozens and hundreds; the basket starfish jerkily closes to a writhing ball of arms and wiggles, accomplishing an entire night's hunting in half a minute.

By day, if you search carefully, you'll find the basket star hidden in a coral crevice, looking like a fist-sized ball of hopelessly tangled yarn. One wonders how it ever untangles that mess to deploy the arms each night.

Another type of nocturnal starfish is the brittle starfish, whose long, slinky arms intrude into many underwater photographs taken at night. If you bring a brittle starfish out of its shelter by day, any number of wrasses, triggerfishes, and other diurnal predators will savagely attack its small central body. Since all of those predators are asleep at night, the brittle starfish emerges, usually draping itself languidly about a sponge.

Leaving the elegant starfish, we come to a family of creatures which may be seen lying motionless on the sand in all the world's tropical seas. These are the placid sea cucumbers, also called sea slugs or *beches de mer*. In some parts of the world they are considered a culinary delicacy, and a major industry collects them.

Sea cucumbers have two major approaches to feeding. One type has an array of tentacles to bring sand and algae to its mouth. These can be fascinating to watch, as one tentacle after another is licked clean. A second group browses along the sand and ingests it directly. Both groups then digest the microscopic crustacea that are in the sand and leave behind them a solid ribbon of compacted sand which resembles nothing so much as a bowl of sandy spaghetti.

If the truth be told, most sea cucumbers are distinctly homely, and I sometimes find myself looking at one and muttering *why*? Still, to the hapless crustacea in the sand these two-foot-long slugs must seem fearsome indeed.

One fascinating sea cucumber, *Holothuria thomasae*, is normally nocturnal and resembles nothing so much as a long vacuum-cleaner hose. It may protrude as much as eight or ten feet from protective coral where it usually anchors its tail. When touched, this pallid hose convulsively contracts, shrinking in the process to a length of no more than eighteen inches.

The final group of slow-moving predators we see in great numbers on tropical reefs are the sea urchins. In some areas of the Caribbean these creatures are seen everywhere, gathering in vast beds of black spines by day, then scattering about the reef by night.

The Caribbean long-spined urchin, *Diadema antillarium*, bears spines up to a foot long, slender and razor-tipped. Unwary divers have filled their fingertips, feet, or knees with these spine-tips, and they can cause excruciating pain. In a way, the common *Diadema antillarium* is the most dangerous marine species to man, because far more people encounter their spines than ever encounter the armament of more exotic predators.

Urchins are echinoderms, scraping algae and crustacea from sand or coral surfaces, processing the mixture, and expelling clean sand from what we see as the top surface of their bodies.

Despite their spiny protection, sea urchins are very subject to predation. I have seen a large triggerfish expel a stream of water to overturn an urchin, then swiftly attack its exposed underside. A broken urchin will attract swarms of damselfishes, wrasses, triggerfishes, porgies, and even moray eels.

Some urchins, such as *Heterocentrotus*, have thick, pencil-like spines. These so-called pencil urchins still hide their vulnerable bodies in crevices during daylight hours and forage only by night.

In the Galápagos Islands, pencil urchins rotate themselves on their spines for a rasping effect to excavate holes in solid granite walls. Some of these walls look shell-pocked, until you realize that each contains an urchin whose spines form an impenetrable barrier to its shelter.

Slow-moving predators such as starfishes are often overlooked in a world of flashing reef fish and larger predators. They deserve a second look, for like the corals, their slow but steady predation methods have stood the test of eons.

Starfish use the strength of their arms to slowly force open the shells of molluscs (Galápagos).

The file shell Lima normally hides its body in coral crevices, using its extended tentacles to detect approaching danger (Belize, Caribbean).

*The nocturnal starfish
Ophidiaster has arms a foot
long, making it a formidable
predator even for larger molluscs (Coral Sea, Australia).*

*Sea urchins rasp algae from
rock or limestone surfaces,
and ingest it through a ventral
mouth (Galápagos).*

Starfish can regenerate an entire body from a severed arm (Maldive Islands).

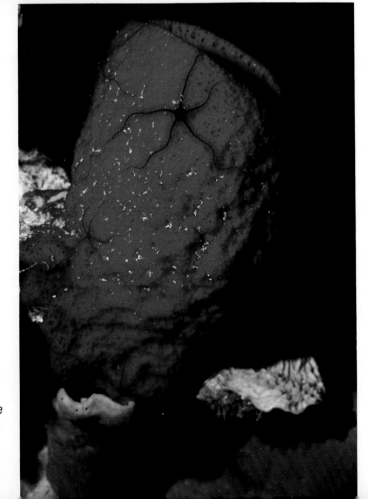

At night the brittle starfish (Ophiothrix) emerges from hiding to feed on the outer surface of a red tube sponge (Grand Cayman, Caribbean).

Few creatures in the sea show greater artistry in their form than the crinoid Comanthina *(Philippines).*

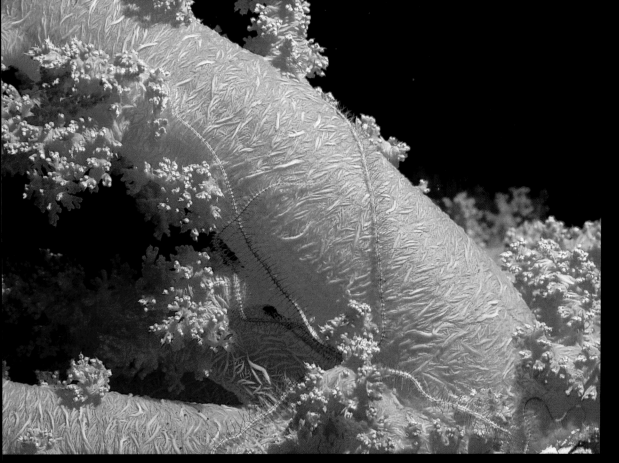

The lacy arms of a brittle starfish (Ophiothrix) entwine about a soft coral (Red Sea).

This is the result of an attack on a coral reef by the crown-of-thorns starfish. The bleached white skeleton left by the starfish is quickly covered by algae. The result is a picture of devastation (Great Barrier Reef, Australia).

For all their plodding existence, starfish such as
Fromia monilis *are some-
times quite decorative
(Red Sea).*

Crinoids (feather starfish) are
found in a veritable rainbow
of colors, including this
blood-red version
(Philippines).

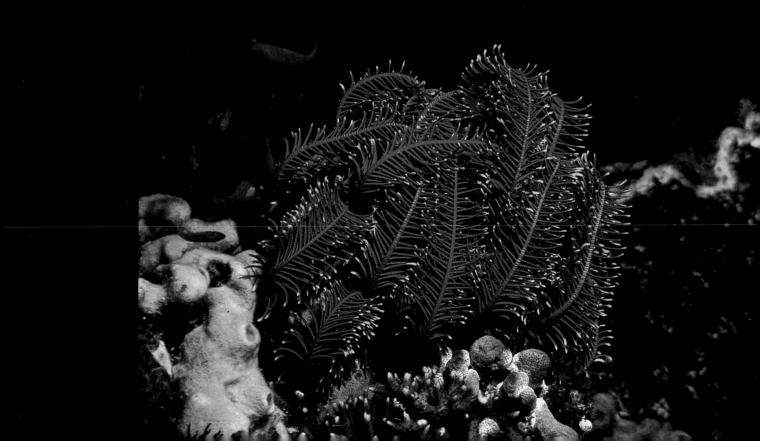

Heterocentrotus mammilatus
*hunts by night, and usually
hides in crevices by day with
its strong spines filling the
entrance (Hawaii).*

*The many-armed starfish
Heliaster is a feature on the
lava walls of the Galápagos
Islands.*

The spectacular sea urchin Asthenosoma *plays host to a pair of tiny shrimp (Philippines).*

3

The Walkers and Stalkers

Most of the predators discussed in our earlier chapters are passive, taking food as it is brought to them by currents or alternatively finding it where it lies on the littered floor of the sea. Now we move on to active predators, who are capable of determined and often swift attacks on their prey.

One unusual group of predators is marked by both their distinctive body structure and their singular method of locomotion. These are the *Arthropoda*, which include the lobsters, crabs, and a host of shrimps. Arthropods are easily distinguished by their hard external skeleton of shell, called the exoskeleton. Their soft body parts are thus protected against minor injury, though we shall later meet several impressive predators on these arthropods which are capable of quickly crushing even these shells.

The external shell has one distinct disadvantage — the soft animal within outgrows it. Thus, these creatures must periodically moult or shed their shells and allow the new one that has grown beneath the exoskeleton to harden. During these periods these crustaceans are uniquely vulnerable.

As might be guessed, one doesn't encounter arthropods out in the open very often. Most are nocturnal, for reasons no doubt echoed by the brittle starfish and sea urchins discussed earlier — that is, if the predators who might eat you are diurnal, you either evolve into a nocturnal feeder or disappear from the scene. Nature's exquisite order has carved out precisely the proper niche for these clanging knights of the reef.

Most sightings of lobsters and crabs in the open occur at night. If you find one by day it is usually hidden under a ledge or in a crevice with only its sensitive antenna-tips showing.

Arthropods scavenge, much as do the slow-moving sea urchins. However, they are also capable of quite determined attack. Some years ago I was working on an underwater movie film; on one of our night dives a small arrow crab (*Stenorhynchus seticornis*) was being filmed. As the powerful movie light blazed, dozens of small wrigglers began to churn in the light-glare. Without hesitation, the arrow crab marched as far out on its coral support as it could and began grabbing wrigglers with both blue-tipped claws. Its predatory instinct was so strong it became absolutely fearless, even in the face of that powerful light-beam.

Similarly, I have come upon huge spiny crabs (*Mithrax*) perched atop dome corals in utter darkness.

The polyps of these corals had their tentacles extended for feeding, and in each case the crab was gobbling up polyps as fast as it could work its mandibles.

While most of us visualize arthropods in terms of lobsters, the family ranges from the great decorator crabs to parasitic crustaceans too small to be seen with the unaided eye. The decorator crab (often called the Sponge crab) is the war wagon among the arthropods; a huge, lumbering behemoth belying the might of its clumsy charge by delicately holding a parasol of camouflaging sponge above its back. The parasol is usually some six inches wide by perhaps four inches tall, pincered into a roughly elliptical shape. It is held in place by two walking legs which have, through evolution, migrated to the rear of the crab's carapace for this special purpose. A first encounter with one of these massive crabs is rather like seeing a National Football League fullback in person for the first time. Alarmed by a diver's bright hand-light, the crab will charge off toward the darkness, bowling over flexible corals and bouncing off the stony ones. Scavenging is a major preoccupation of these crabs, and it is easy to imagine other scavengers backing off when this monster lumbers upon the scene.

Other smaller crabs emerge at night on coral reefs from the Caribbean to Australia to the Red Sea, some scavenging on open sand (though never far from shelter) while others pursue the feeding coral polyps. Since large groupers and nurse sharks, among others, prey on these crustaceans, it's no wonder that they are easily alarmed and always ready for flight.

The lobster family is familiar to us from our dinner plate and pictures of graceful, long-legged *Panulirus argus* or *versicolor*. There are few joys to match the discovery of a hunting lobster abroad at night. The joy is multi-faceted, partly the appreciation of the creature's armored elegance, partly tingling memories of past meals, and partly an intense desire to see just what it might be eating.

While we are all familiar with the long-legged lobsters (*Panulirus*), there is another branch of the family with a vastly different appearance. These are the slipper lobsters, which look more like battle tanks than reef creatures. Their shells are flattened into flared skirts at their edges, and their short legs are hardly visible beneath the hard carapace. Slipper lobsters are principally scavengers, and night divers will discover them motionlessly feeding on open rock or amid coral. Sometimes because they are motionless we may draw the erroneous conclusion that they are incapable of rapid motion.

Quite the contrary. One starry night I was prowling a darkling reef far out in Australia's Coral Sea. These reefs are 200 miles from shore, far beyond the outer limit of the Great Barrier Reef, and you can run into anything out there.

Without warning, I suddenly found myself very close to a very large slipper lobster. In numerous past encounters I had never seen one move rapidly, so I settled down with my camera to take a picture. As I bored in with the camera there was a sudden crash, and the lobster disappeared from view. Instinctively I swung the light around, to discover the lobster 30 feet away and 10 feet above the reef, moving like a rocket. It settled to the coral again and began feeding in the distance as if I had never intruded. I left with a far greater respect for these unusual animals.

The most graceful of the arthropods are the amazingly varied shrimplike crustaceans. These range from the open-ocean krill of the Arctic and Antarctic regions, principal food of whales, to the delicate, inch-long *Periclemenes* shrimp perched upon the anemones or corals of a tropical reef. Shrimps, like their larger cousins the lobsters and crabs, are succulent delicacies for a host of fish. For this reason they are most often seen hidden in crevices with only their long, graceful antennae protruding. This is particularly true of the larger reef shrimp such as the barberpole shrimp, *Stenopus*, and the peppermint shrimp, *Hippolysmata*. In our chapter on "odd couples" we'll find that these shrimp perform certain specific cleaning services for large fish and eels. It is important to point out, however, that when these creatures are not specifically protected by the cleaning ritual, they are gobbled up in a minute if exposed.

One fascinating aspect of the predatory life of shrimps is that they are prime carnivores of their smaller relatives. Tiny isopods, copepods, and zooplankton are all grist for the shrimps' mill, and their grace and defenselessness should not blind us to the fact that from the copepod's point of view they are a fearsome apparition.

Rather than hiding in crevices, small shrimp such as *Periclemenes* may adopt a totally different survival strategy. *Periclemenes pedersoni* daintily walks the seething tentacles of Bubble Anemones *Heteractis* impervious to the stinging nematocysts of the anemone and protected by them. Any fish which darts in for a bite may find itself devoured by the anemone. A similar relationship is pursued by *Periclemenes yucatanicus* with different species of anemone, *Condylactis.*

In the Red Sea, another species of *Periclemenes* seems equally at home on Bubble Anemones or coral. In all of these cases, the anemone's lethal reputation serves as excellent protection for the highly vulnerable shrimp.

We could not end any treatment of arthropods without a close look at *Squilla*, the mantis shrimp. For its size, this is one of the most awesome predators in the sea. If *Squilla* measured six feet instead of six inches long, scuba diving would have been a short-lived sport indeed.

There are at least 20 species of mantis shrimp, and one can learn a lot about the way they operate simply by hearing their colloquial name: "Thumbsplitter."

These predatory crustaceans live in tunnels built in open sand, or in crevices in rocks. Sometimes their bodies are poised at the mouths of their burrows, watching the passing parade for possible prey. Other times the burrow will stand empty — but approach it closely and you are confronted by a raging mantis shrimp ready for battle.

The mantis shrimp has an elongate, lobster-like body, but its head and jaws are something out of a Hollywood monster epic. The second pair of thoracic legs have evolved into weapons similar to the arms of the terrestrial praying mantis. However, where the arms of the praying mantis hinge at the top, the arms of the mantis shrimp invert that structure and put the hinge at the bottom. Otherwise the attack of the two creatures is similar. An upward strike — is too quick for the eye to follow — and the prey is dispatched, often sliced cleanly in half.

Once, while trying to take pictures of this prodigious predator, I hung bits of fish above it on a hand held fishing line. Then I tried to release the shutter at the moment the shrimp struck. Despite infinite (and uncharacteristic) patience, the entire series of photographs reveal only a cloud of sand, and the mantis shrimp back in its burrow with the bait. Score: Photographer zero, shrimp about ten.

Perhaps the most fascinating feature of this aggressive shrimp are the prominent and unique eyes. These are on stalks, and look like Mexican jumping beans. The shrimp can rotate them on the stalks, an action which is totally disconcerting to the onlooker. Moreover, within the moving beans are tapestries of migrating flecks of black and white. As I said, if this

ferocious visage and prodigious armament were not executed in miniature, the coral reef would be a hazardous place indeed. In actuality, *Squilla* is seldom encountered by divers anywhere in the world.

In reviewing the arthropods, I've concentrated on those most readily visible, which are, naturally enough, the larger members of the family. By far the greatest numbers of crustaceans in the sea are rarely seen. The curious diver may catch a glimpse of these smaller crustaceans in several ways.

One way is to peer closely to an open, sandy bottom. While at first glance this terrain looks empty and barren, a second look may be a shock. The sand is literally alive with miniature crustaceans and tiny fish, all colorless to blend into the pale sea floor. Beneath the sand, razor-fishes, molluscs, mantis shrimp, jawfishes, eels, and other creatures burrow unseen, all eager to partake of the sandy feast. From above, fish such as goatfish, trunkfish, wrasse, and triggerfish will stir the bottom, churning up the crustaceans in a feeding pattern that is repeated countless times each day.

A second place to see certain parasitic isopods and copepods is on fish. Groupers, soldierfishes, moray eels, and certain snappers seem particularly afflicted. On a darkened grouper you can just barely make out a dozen or more skittering spots moving about the surface. Later, in our chapter on "odd couples," we'll examine the complex cleaning phenomenon and see these creatures more closely.

A third place to reliably see a profusion of small crustaceans on the reef is on the body and arms of any crinoid (feather starfish). Each crinoid may have several hundred isopods or copepods cadging food from the highly efficient food-gathering system described in chapter 2.

Finally, whenever you look about you in the green waters off the Galápagos or California or any cold-water area, the lack of clarity is largely due to plankton. These plankton are the true miniatures of the crustacean world. They are the thin soup that nourishes wide-mouthed manta rays, whale sharks, and true whales as they move in the open sea.

From all this you can see that the crustaceans are among the most numerous of all the sea's families. If you know where to look, you can find them in incredible abundance. With the coral polyps and the phytoplankton, the crustaceans constitute the broad base of the sea's majestic food chain.

Stenopus hispidus, *the bar-ber-shop cleaner shrimp, carries its young on its back (Maldive Islands).*

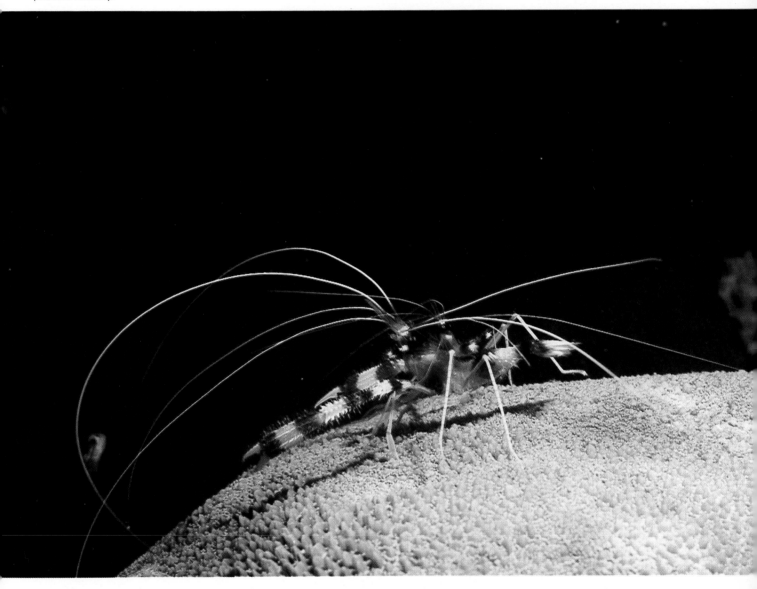

The Harlequin shrimp (Hymenocerus picta) preys on these red starfish (Hawaii).

The painted crayfish or lobster (Panulirus versicolor) can be seen by day invariably beneath coral ledges with loose sand bottoms (Philippines).

Despite their impressive armor, many crabs are abroad only at night (Philippines).

The Indian lobster (Panulirus guttatus) *is richly colored, but hides deep in coral crevices during daylight. It is a nocturnal feeder (Fiji).* ▶

These brilliantly daubed shrimps hide in rocky crevices, providing occasional cleaner services to large green moray eels (Galápagos).

The red-bellied batfish (Ogcocephalus darwini) *walks about the volcanic sand bottom searching for small crustaceans (Galápagos).*

The burrowing mantis shrimp (Squilla) *is an intensely aggressive predator, and is known colloquially as "Thumbsplitter" (Bonaire, Caribbean).*

These huge Mithrax *crabs come out from their coral shelters to pluck coral polyps which are exposed during nocturnal feeding (Belize, Caribbean).*

These massive decorator crabs have specially evolved legs to hold sheltering parasols of sponge above their backs (Palau).

The cleaner shrimp Stenopus hispidus *offers its services from a colorful sponge (Belize, Caribbean).*

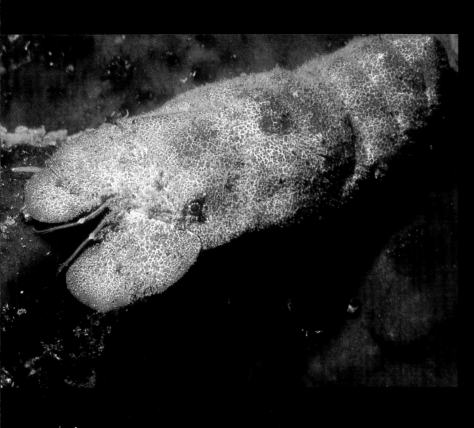

These large slipper lobsters (Scyllarides) occur in great numbers on the rocky flanks of the Enchanted Islands (Galápagos).

Small Periclemenes shrimp move unconcernedly amid the nemotocyst-armed tentacles of anemones (Fiji).

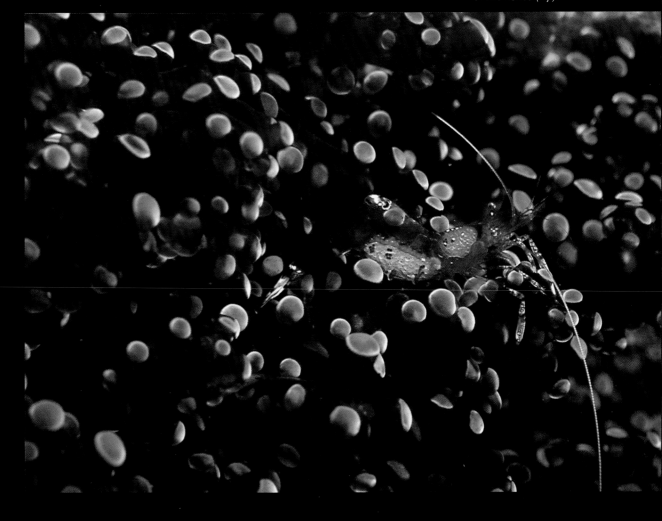

4

Creatures That Glide

Some of the most placidly colorful and flashily dramatic of the reef predators are legless and (at least in part) finless. Simplest of these creatures are the marine flatworms. Flatworms are often brilliantly colored on their exposed backs. Their white undersides are smooth, and they have the ability to adhere to solid surfaces. Flatworms will be found most often gliding along dead, algae-covered coral skeletons. Here they browse on the algae, seemingly oblivious to the busy reef about them.

The reason you won't find them on live corals is that the corals sting their sensitive undersides. On those rare occasions·when I have attempted to move a flatworm from an ugly dead coral surface to a beautiful live one, the flatworms have curled up and allowed themselves to fall free in the water, looking no doubt for a more hospitable location.

I've seldom seen flatworms in the Caribbean but they are abundant on coral reefs of the South Pacific. They range from one inch to perhaps five inches in length, and occur in pink, burgundy, yellow, green, and several combinations of black and white.

Besides adhering to coral surfaces, flatworms are also capable of an undulating, rhythmic swimming motion in open water. On a few occasions I have encountered them swimming several feet above the reef, apparently of their own volition, perhaps using a mild current to find new feeding areas.

On one dive in New Caledonia I was privileged to see an apparent sexual encounter between two flatworms (*Pseudoceros*). The two worms came from opposite directions, browsing on algae until they were perhaps two inches apart. Then both worms reared up, exposing their white undersides to each other. On the underside of each worm quickly developed what looked like female breasts; when these organs had fully extended, the two worms seemed to dance facing each other. For several seconds they swayed like miniature cobras, their extended organs seeming to grope for each other.

Then, as quickly as the erotic dance had begun, it ended. Both worms retracted their protruberances, lowered themselves to their regular gliding positions, and moved away from each other. It was as if the encounter had not even occurred, both worms browsing contentedly along in the afterglow of their brief affair.

Another vivid family of gliders is that of the nudibranchs. These shell-less snails are among the most extravagantly colorful of all reef-dwellers, for reasons which are not really well understood.

Like the flatworms, the nudibranchs are possessed of a prehensile underbelly. This ability to adhere to any surface means that you may find nudibranchs upside down on the roofs of undersea caves or in other spots protected from direct sunlight. Many nudibranch species are nocturnal, including many of the largest specimens such as the gaudy scarlet Spanish dancer, *Hexabranchus sanguineus*. Still, you may often encounter nudibranchs by day, especially on deeper reefs at 100-foot depths or more, on shadowed walls or in crevices.

While most nudibranchs are small (one to three inches), some larger species can reach six to twelve inches in length.

There are many kinds of nudibranchs. Some have rows of fingerlike projections on their backs, others are grooved. Still others, the dorids, have large, distinct tufts of gills on their backs. In many gaudy species such as *Chromodoris* and *Notodoris*, these external gills form a gaily-hued plume mounted atop the nudibranch's back.

Some species of nudibranch cheerfully prey on those corals, anemones, and hydroids which, as mentioned earlier, are armed with chemically activated stinging structures called nematocysts. These nudibranch species manage not only to ingest the lethal nematocysts, but also to suppress their firing mechanism.

The ingested nematocysts migrate through the nudibranch's system, making their way to the external gills, where they remain. This secondhand armament certainly explains how those species of dorids deter predators. The word has gone out in an evolutionary manner that taking a bite of these nudibranchs earns you a mouth full of pain.

There would seem to be no intraspecies purpose to the brilliant body-colors of the nudibranchs. This leaves open the possibility that the spectacular display is to clearly identify the animal to potential predators. With their reputation as a dangerous prey, they avoid being bitten by mistake — by being unmistakable.

Nudibranchs, like flatworms, don't appreciate being set down on live corals. They will curl up into a ball and let the water carry them away, uncurling only when they are free of the stinging tentacles.

Both nudibranchs and flatworms are the most placid of reef-dwellers, slowly gliding their way across the reef in one long, unhurried meal. Often divers fail to spot these brightly hued animals because they are not moving or are feeding in the shallows. In most caverns or intricate reef structures are plentiful nudibranchs, and some patient searching will often yield extravagantly colored specimens. They may also be a possible reward for your patience in examining thickets of dead coral skeletons.

While most dorids are nearly motionless as they feed, at least one type nearly dances. Every second or two the lovely *Chromodoris coi* flares its mantle, almost like the motion of a manta ray's wings. Often that's how, out of the corner of your eye, you detect this pulsing motion and discover this spectacular creature.

Many other molluscs pursue a similar lifestyle to that of the flatworms and nudibranchs, though with the solid protection of a hard external shell, which they continue to build ever larger by the secretion of limestone.

Cowries hide their bodies within their characteristic shells, but extend a soft, fleshy mantle to completely cover the shell's outer surface. The egg cowrie (*Ovula ovum*) of the Pacific covers its snow-white shell with a jet-black mantle decorated with small white spots. The tiger cowrie (*Cypraea tigris*), on the other hand, covers its lushly spotted shell with a muted gray mantle, and the mantle is covered with erect protrusions which may act as feelers to alert the cowrie when is entering a narrow space. The familiar flamingo tongue (*Cyphoma gibbosum*) of the Caribbean completely covers its bland shell with a florid, spotted mantle.

One profoundly impressive mollusc is the triton trumpet, *Charonia tritonis*, which was mentioned earlier as a major predator on the crown of thorns starfish *Acanthaster*. The triton trumpet displays a kind of muscular control with its rather clumsy-looking foot that is extraordinary. I once observed this mollusc mold its foot about the sharp spines of a sea urchin, grasp the spine-tips, and lift the urchin off the coral. Thus deprived of motion, the urchin was helpless. Slowly, the triton trumpet rotated the urchin spine by spine until its unprotected underside was revealed. Then it plunged its tubular proboscis into the urchin and sucked out its tissues.

This demonstration of technique shows how easy it is for us to be led astray by external appearances. Most observers watching *Charonia* slowly glide over the coral surface would think it a plant-eater. How could anything so peaceful ever attack anything?

Then observe the triton as it senses the urchin prey nearby. It rears its body upward, lifting its shell high into the air. Turning, it locates the urchin and glides swiftly forward, simultaneously forming the forward end of its gliding foot into a grasping protrusion. When the spines of the urchin touch this extended fleshly foot they seem to penetrate and stick; the urchin is doomed. The urchin, whom we saw earlier as a predator, has abruptly become prey.

Moving to a different class of gliding animals, we find the eels. Legless, and in most cases lacking any pectoral fins as well, the eels nevertheless are superb predators. One mistake divers often make is to think of only large eels such as the green moray (*Gymnothorax funebris*) as predators. There are in fact much smaller eels which are far more aggressive than the moray. One is a foot-long Indian Ocean species, which I call the "tin-snip" moray because its lower jaw is as large as its upper jaw, giving it the look of a pair of tin-snips.

I've had these little demons nip me numerous times. In the Maldive Islands, as I was photographing one near a coral head, another bit my hand. I soon discovered that a half-dozen or more swarmed beneath that one coral head, and they had no hesitation about biting any hands, knees, or even rubber fins that came within range.

Another aggressive small moray is the viper moray (*Enchelycore nigricans*), a fearsome-looking eel with long, curved jaws that seldom exceeds six inches in length. Good thing, too — unless you're a small fish or crustacean living in a narrow crevice in the coral reef. Though you may be safe from most other creatures, these small morays can insinuate themselves into holes so that they literally thread the reef in their search for food.

In rocky areas such as the Galápagos, huge morays lie right out in the open in granite cracks, while in New Zealand huge burnished-bronze morays are everywhere. Break a single sea urchin, and two or more of these aggressive eels will attack it savagely. It is amazing to see their antics as they contort their bodies in order to get their jaws inside the shell.

Speaking of contortions, an amazing demonstration occurs when a moray encounters something too large to swallow. First it may cleverly try to wedge it in a small coral opening so it can tear off a chunk. Failing that, it will tie its body in a knot, pull its head through the loop and use its knotted body as "hands" to tear out its bite.

Another incident may demonstrate the astonishing versatility of eels as predators. On a dive in the Caribbean I chanced upon a native fish trap of wood frame and chicken-wire. Two aggressive large-mouthed eels of a reclusive type I had never before seen had entered through the holes in the wire.

These eels had immense mouths for their body size and had swallowed small fish larger than their own girth. I could clearly see the outlines of the swallowed fish swelling their bodies. Now the eels were trapped. With these oversized fish in their bodies they could no longer fit themselves through the mesh.

As I approached the cage the eels sought to escape. Each eel would put its head through one hole in the cage and its tail through another, but the ingested fish stopped them each time. As I came closer one eel, showing alarm, convulsed its body, opened its huge maw and regurgitated the partially digested fish carcass. Then it swiftly slipped through the mesh and disappeared into the reef, leaving its mate stranded.

The second eel, apparently in a triumph of appetite over fear, remained in the cage to digest its meal.

Amazing vignettes of eel behavior are surely not limited to morays. On one occasion I came upon a sharptailed eel (*Myrichthys acuminatus*), completely motionless on the open reef. Thinking it wounded or dying, I photographed it for several minutes at point-blank range. No wounds were visible, and I could see no reason for its total torpor. After I finished and was about to leave with my heart full of sympathy, I was taken aback to see the eel abruptly recover, look about, and slither away through the corals. Had it been asleep? Or perhaps digesting a meal? Whatever its reason, it would not thereafter let me approach as closely again.

One other type of reef-dweller could be included here, among the gliders, or in later chapters on swimmers, for it is capable of both methods of locomotion. This is the largely misunderstood family of *Hydrophiidi*, the air-breathing sea snakes. All manner of myth and calumny has grown up about these creatures, depicting them as savage and bloodthirsty. Nothing could be further from the truth.

These are true snakes, relatives of the notorious land cobra and armed with even more powerful venom. Where morays have at least a dorsal fin if not pectoral fins to guide them in swimming, the hydrophids have no fins at all. Instead, they have evolved a distinctive paddle tail which they utilize with impressive efficiency to guide themselves through the water.

By day, sea snakes are normally found curled in hoops on the reef, sleeping. Some species in the Philippines are communal, with six or more snakes writhing into a single coral crevice, looking for all the world like a bowl of striped spaghetti.

I said earlier that these marvelous predators are largely misunderstood, and that is certainly true. When I take groups of divers to Australia and the Philippines, many have heard wild tales about the snakes. They enter the water with great trepidation, looking all about them to spot the first snake. Within two or three days these divers are kicking snakes out of their way in order to photograph other animals. How do you reconcile myth with reality? Is the snake a frightening predator or a pussycat?

The answer is, "both, depending on your point of view." These hydrophids are extremely efficient predators, poking their slender heads by night into the myriad coral crevices where small reef fish seek refuge. In the darkness the small fish are totally helpless.

The snakes, however, have very fragile jaws, which could be broken by a struggling fish. Now we see the reason for the snakes' powerful toxin. They quickly inject the fish with fast-acting venom; moments later, they can swallow the paralyzed prey without a struggle. Moreover, the toxin is so powerful it later digests the fish. These sea snakes have no digestive system *per se*.

From the human point of view the snakes present no threat unless they are netted, trapped, cornered, or injured. Why would they gratuitously bite an immense human they could not possibly swallow? My decade of diving with hundreds of snakes of seven different species confirms my feeling that unless in desperation they will not bite humans.

On the contrary, several Australian species seem to find us great curiosities, and spend long periods of time studying us closely. They particularly like the turbulence behind our swim fins, and will often be seen tumbling like happy puppies in the backwash from a diver's fin-strokes.

Moreover, they find our gleaming equipment irresistible. Our shiny metal and glass cameras, our wrist instruments, and even the glass of our face-masks are objects of intense scrutiny for these gentle and essentially harmless creatures. I confess, though, that the first time a five-foot olive snake (*Aepysurus laevis*) starts darting its forked tongue about the glass an inch from your eyes it can be a trifle disconcerting.

What you do is stay still and let the snake investigate. Rapid movement might frighten it, which is precisely the wrong thing to do. If you wish to handle the snakes, the balance point is a few inches behind the head, perhaps one-quarter of the way along the body length. Hold them lightly — don't squeeze. Usually they will accept gentle handling very well. If a held snake wants to leave, you'll feel a powerful flex of its muscles, and it will back out of your grasp, sometimes tonguing your hand as it leaves.

Sea snakes are air breathers, with a long air bladder in their bodies. Depending on their level of activity, their air supply will last at least a half-hour, but no more than two hours. Then they will make a long, sinuous ascent to the surface. Lolling on the surface for perhaps a half-minute, they will ventilate and refill their air bladder, then undulate back down to the bottom. Except for these forays to the surface, sea snakes are bottom-gliders, ever searching for the next small prey.

Whatever their morphology or prey, these creatures I have characterized as gliders exhibit a special gracefulness as they hunt. Seldom moving swiftly, they come upon their prey with a kind of deliberate, ineluctable intensity. They are the irresistable force in their world.

Another characteristic they share with other gliders, though for apparently different reasons, is a lack of creatures preying upon them. Nudibranchs are known to be both distasteful to fish and often armed with second-hand nematocysts. Morays seem to be the kings of the crevices, and avoid moving about in the open away from their coral mazes. As for sea snakes, I have witnessed them swimming right into the middle of a shark frenzy to take some of the bait. Not one shark gave any hint of interest, so the word seems to be out not to tangle with these lethally armed reptiles.

The reef is certainly enriched visually by the presence of these wondrous citizens, and their predation, while not showy, is most certainly efficient.

The gentle-looking triton trumpet (Charonia tritonis) displays its martial skills by plunging its proboscis into a captured sea urchin (Curaçao, Caribbean).

In the darkness below 125-foot depths, the nudibranch Chromodoris feeds on a sponge whose fiery color is revealed only under artificial light (Philippines).

The volute shell Voluta *glides easily across a coral head, its siphon held high (Coral Sea, Australia).*

Moray eels can extend their bodies far from the protection of the coral reef if they choose (Maldive Islands).

Sea snakes can grow to great lengths, but must breathe air to survive. Here the diver accompanies it to the surface (Philippines).

I discovered this six-inch flatworm browsing on dead coral rubble at a depth of 175 feet (Fiji).

This longhorn nudibranch
Tambja *is shown browsing
on a cluster of tunicates
(Philippines).*

The sea snake Aepysurus
duboisii *studied me very in-
tently as I approached (Coral
Sea, Australia).*

Rhinomuraena, *the rhino moray or ribbon eel, may warn you away from its hole with its aggressive manner (Philippines).*

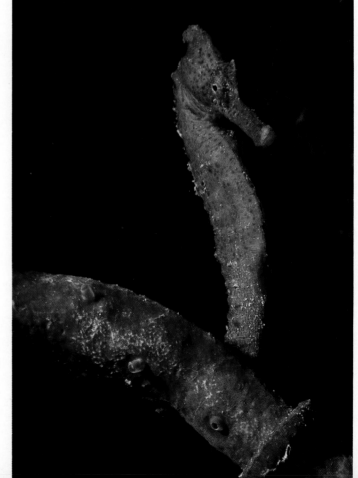

Its hooked tail enables the seahorse Hippocampus ingens *to attach itself to sponges, gorgonians, or even projecting metal (Virgin Islands, Caribbean).*

The nudibranch Notodoris pulses the edge of its body every second or two, giving it an air of bustle (Fiji).

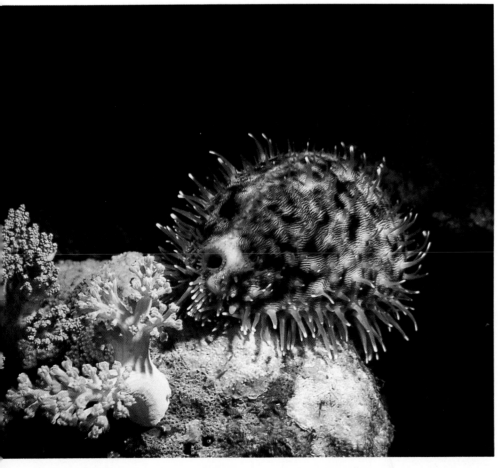

Tiger cowries (Cypraea tigris) spend their days in shadowy crevices, but emerge at night in their formal mantles (Philippines).

This colorful nudibranch is able to adhere itself to any surface. Here it finds its prey on a coral (Philippines). ▶

This nudibranch Pleuro-
branchus *is nearly a foot
long, and occurs in every
conceivable combination of
white and burgundy red
(Coral Sea, Australia).*

*The large nudibranchs known
as Spanish dancers weave
sinuously when swimming in
open water. This is* Hexabranchus imperialis *(Papua
New Guinea).*

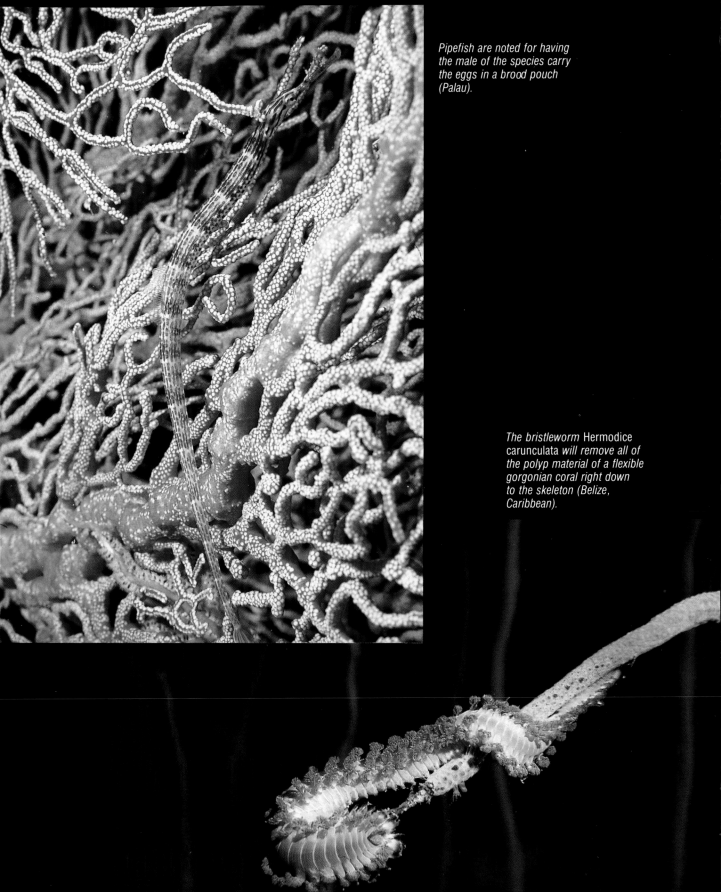

Pipefish are noted for having the male of the species carry the eggs in a brood pouch (Palau).

The bristleworm Hermodice carunculata *will remove all of the polyp material of a flexible gorgonian coral right down to the skeleton (Belize, Caribbean).*

5

The Swimming Browsers

This group of predators includes many of the tropical reef fish, who are at the same time the most familiar reef-dwellers and the easiest for us to befriend.

That, of course, is because fish have been familiar to us since childhood, while other reef creatures may seem new and foreign to us. Reef fish also project a certain amount of personality, their tiny bright eyes sparkling at us as they pass. Finally, we feel we understand a fish better than we do a nudibranch or a coral polyp. After all, we swim around, too, using our hands and feet as fins, and looking at things about us with our own bright eyes.

All of this whipped-cream perception may blind us to the realization that these browsing fish are in fact deadly predators to whole classes of creatures below them in the food chain. Those bright-eyed faces are in fact focused with great pragmatism on their next snack, and they reserve their brightest smiles for us when we come with food in our hands.

The swimming browsers naturally bifurcate into two general groups depending upon the food they eat.

The first group I would call the plankton-pluckers, and even this group further subdivides into those who pluck their plankton within inches of shelter, and those who school several feet above the reef but seldom stray into open blue water.

The second major category is the bottom-browsers. These graze on corals, algae, tiny crustaceans in the sand, sponges, and other prey that is actually part of the reef rather than the current-borne food stream.

Taking the plankton-pluckers first, I use the word "pluckers" quite precisely for its visual imagery. These fish are usually small, often brightly colored, and include such species as the grammistids and fairy basslets. What is unusual about these dancing, darting splashes of living color is that at the first sign of larger predators many of them disappear like a flash into their coral burrows. This group includes *Gramma loreto*, the royal gramma of the Caribbean; *Aspidontus taeniatus*, the saber-toothed blenny; and the varied species of *Nemateleotris* in the Indo-Pacific. Usually these burrow-divers occur singly or in pairs, though in some teleotrids a dozen or more may share quarters. The saber-toothed blennies have their own specialty, diving *tail-first* into abandoned serpulid-worm tubes, whereupon their tiny heads reappear bobbing at the mouth of the tube waiting for an opportunity to emerge and feed once again.

Fairy basslets such as *Anthias*, on the other hand,

may occur in hundreds or thousands, a cloud of tiny orange or purple fish confusing would-be predators with their sheer numbers.

All of these species soar and dance above the solid coral, plucking planktonic tidbits from the passing food stream. Below them are the rooted corals, walkers, and gliders, above them the dangerous open blue water.

The fairy basslets have one fascinating distinction. In a school of dozens or hundreds of individuals there will be but a single male, always of a different color. In the gaily orange *Anthias squammipinnus*, the crowding females with all be orange, while the solitary male will be intense purple.

When any harm befalls the male, its place is taken within hours by a female which changes both sex and color. This is one of the wonders of the sea's endless tapestry of life.

One type of anomalous plankton-plucker which eschews the ignominy of burrows is typified by several species of *Caesio*, the fusiliers. These graceful fish form long, strung-out schools which follow the contour of the reef surface at an elevation of 5 to 20 feet. In the late afternoon the school will scatter, and the fusiliers will fill the open water perhaps 20 to 50 feet from the coral surface, dancing in a predatory frenzy that fills the light-fading water with motion.

One other completely fascinating plankton-plucker is really more like a plankton devourer. This is the schooling mullet (*Mugil*). In a fascinating evolutionary adaptation, this usually dish-shaped fish can suddenly balloon its jaws into huge scoops perhaps five times its normal frontal area. These colossal maws scoop in plankton for a second or two before closing again as the mullet swallows its meal. Watching these unique gulpers one almost imagines them shouting as the great scoops open.

From the plankton-pluckers we move on to the surface browsers, a much larger and more variegated spectrum of predators.

The most familiar of the browsers would certainly be the ubiquitous butterflyfish and angelfish, and the parrotfish. While these familiar creatures might seem to use similar predation techniques, they are in fact as different as can be.

The *Chaetodontidae* (butterflyfishes) are to the fish world what the soft corals are to the coral world — lovely, brilliantly plumed aristocrats. Capable of a rainbow of exquisite finery, they stand out among all their more humdrum brethren. Our eye is instantly drawn to these elfin courtiers among all other fish.

For all their visibility and fascination, the precise reason for their distinctive coloration has long been a bone of contention among marine scientists. The distinguished psychologist Konrad Lorenz opened his seminal volume *On Aggression* with a treatise based on his personal observation. To Lorenz, the color patterns were for easy intra-species identification to avoid what would otherwise be inevitable conflict over turf, food supply, or even breeding partners.

Recently, other scientists, notably Paul Ehrlich, have taken issue with Lorenz's conclusions, citing aggressive behavior by drably colored damselfishes. Someday perhaps we'll know the real reasons, but while the research goes on we can simply enjoy these dainty predators on their obvious visual merits.

The angelfish and butterflyfish are usually placed together in the family *Chaetodontidae*, though some scientists insist that based on morphological evidence they may not even be closely related. It is certain that they share some physical characteristics: disc-like bodies, small mouths, and the comblike teeth which give the group its name (*Chaetodontidae* is derived from the Greek for "comb-tooth" or "bristle-tooth").

Angelfishes have preopercular spines, starting in front of the gill cover, which immediately enable an observer to identify them; size is definitely not a guide, though new divers who have only experienced the Caribbean may conclude that the butterflyfishes are the small ones and the angelfishes are the big ones. In the Indo-Pacific you will find several small angelfishes such as *Euxphipops trimaculatus* or the cherubfishes (*Centropyge*), and huge butterflyfishes such as *Anisochaetodon lineolatus*.

In the butterflyfishes there is a strong tendency toward color bars or body-stripes, especially eye-stripes, in most species. Some species are strongly striped as juveniles, and in that phase of their life offer cleaning services to larger fish (see chapter 7). Another interesting feature is that some species which are not banded by day develop clearly banded patterns at night (see chapter 11).

The predation of angelfishes and butterflyfishes often appears quite dainty, but is nonetheless extremely efficient. Some species of these fish eat coral polyps, other sponges, worms, and tiny crustaceans; some are omnivorous.

One clue to their eating habits may occur in the shape of their mouths, as in *Forcipiger*, the clear winner of the Cyrano de Bergerac award for distinguished nasal development. The extended mouth of this exquisite fish may be thrust forward nearly half a body-length in front of its eyes. I have watched these superb predators hover precisely before a coral head, delicately adjust their body position with their active pectoral fins, then dart forward and pluck coral polyps with the accuracy and deftness of a surgeon.

One clear advantage of this long snout is to protect the butterflyfish's eyes from damage as it feeds, particularly in surging shallow waters. Not only *Forcipiger* but numerous other chaetodontids have protruding mouths; *Forcipiger*'s is merely the longest in relation to body length.

Other butterflyfishes and most angelfishes, by contrast, have rather blunt, rounded snouts. They will be seen nipping chunks out of sponges or nosing amid the stirred crustaceans churned up by any disturbance which roils the sandy bottom.

Other families share the extended-snout design of the long-nosed butterflyfish. One of the most fascinating is that of the seahorses and pipefishes. With their tiny horse-shaped faces, these unusual fish browse on coral polyps with precise dartings of their head. Their extended snouts serve, as in the chaetodontids, to protect their eyes from the lashing tentacles and hard skeletons of the corals.

Parenthetically, the loss of an eye in many fishes is a death warrant, for predators will soon take the fish from that blind side. I have on several occasions seen fish with damaged eyes hover with their blind side to a stony wall as if to see danger approaching.

While seahorses (*Hippocampus*) have the familiar vertical stance and hold themselves in place by wrapping their prehensile tails about corals and sponges, pipefishes by contrast are long, slender, and eel-like. Pipefishes (*Syngnathus*) are often confused with small eels by new divers. Beside its miniature horse-face, another distinctive feature of the pipefish is a round caudal fin the size of the fingernail on your little finger; this caudal is often flared and usually pink in color.

Pipefishes and seahorses have long been famous for their distinctive method of protecting their eggs until hatched. In all but one species, syngnathid males carry the eggs in a brood pouch. The odd species in which the female carries the eggs is *Solenostomus paradoxus*, the beautiful ghost pipefish. This ornately decorated pipefish is adapted to resemble the complex feeding arm of a crinoid. Indeed, when I first discovered *Solenostomus*, I thought it a misshapen arm of the

crinoid I was about to photograph. When the "misshapen arm" darted away from my fingers I realized what this unique creature must be.

Other common swimming browsers on the reef include triggerfishes and their cousins the filefishes, trunkfishes, puffers, and porcupinefishes. All of these very different-looking browsers are plectognaths, grouped because of their spinous dorsal fins. I'll treat the trunkfishes, puffers, and porcupines in our chapter on defense (chapter 9), but the triggerfish and filefish may be included here among the browsers. Both the triggerfish (*Balistidae*) and the filefish (*Monocanthidae* and *Aluteridae*) are immediately recognizable by the dorsal spine they erect when threatened or angry. In the balistids this dorsal spine is used in a very special way: If approached too closely, the triggerfish will dive into its coral burrow, erect its spine and lock itself in place. You can grasp its often-exposed tail and be unable to pull it out.

Monocanthids (filefishes) are smaller, and tend to hide among the supple arms of gorgonians. Larger aluterids such as the scrawled filefish, *Alutera scripta* , are reduced to swimming away from potential predators.

The balistids strike the experienced observer as quite clever and can be extremely aggressive. Large individuals of *Balistoides viridens* will lift huge chunks of coral in their teeth to uncover the brittle starfish, sea urchins, and small crustaceans beneath. Entire feeding frenzies are started by this fish I call "Charley the Wrecker," as angelfishes, wrasses, butterflyfishes, and others crowd around to share the spoils. Charley is the boss, though, and the other fish keep out of his way. Similarly, I observed the stunningly ornamented clown triggerfish, *Balistoides conspicullum*, hang back and watch while smaller fish took my proffered fish-bait. When convinced there was no danger, this peacock of the fish world charged through the pack, brushing aside all the other fish to be first in line for the food.

There is one predation technique shared by these large triggerfishes and their cousins the trunkfishes: They are capable of blowing powerful jets of water. In most cases, these firehose bursts are used to churn up the sandy bottom to reveal small crustaceans or molluscs. In one fascinating variation, a large triggerfish will fire its jet with great precision and lift a sea urchin clear of the bottom, simultaneously tipping it over. Then the triggerfish charges in to attack the unprotected underside, demolishing the hapless echinoderm in seconds.

Another family of spined browsers are the surgeonfishes (*Acanthuridae*). These pancake-shaped fishes range in color from blue or gray to canary yellow. Unlike most of the browsers we've discussed thus far, the surgeonfishes are herbivorous, browsing on algae. Since most dead coral or sunken metal is quickly coated with marine algae, the surgeonfishes are never at a loss for food.

Surgeonfishes are specifically identified by the sharp, extensible spines on either side of their caudal peduncle (the base of the tail). These spines can be razor sharp and formidable, and are clearly a deterrent to would-be predators.

Surgeonfishes will often gather in dense schools and on occasion will "swarm." In this swarming behavior the entire school feeds simultaneously in a densely packed, churning mass. This is the only time these fish ever seem other than totally serene.

Many fish have pelagic larval stages in which they may drift in ocean currents for hundreds or thousands of miles before settling out onto a new reef; the acanthurids have one of the longest-lasting larval stages of all. Early biologists even thought this larval stage was an entirely different species of fish until its true nature was discovered.

Another family of fish in which a similar error was made is the wrasse (*Labridae*). The Pacific wrasse, *Bodianus axillaris*, is roan-colored as an adult; in its juvenile stage, however, it is spectacular jet-black or maroon with bold white spots. Early taxonomists thought the small, white-spotted stage represented a separate species. Indeed, the immense variability of labrids leads some scientists to speculate that future research will reduce the present number of species by half. Many wrasses change color and/or sex as they mature, and certain males (supermales) are different in coloration from other mature males. In addition, some wrasses (for example, the razorfish) lose the spinous dorsal fin as they mature. All in all, there is still much confusion as to the precise taxonomy of these fish.

The labrids range in size from swarming miniatures such as *Doratonotus*, dwarf wrasse, to the immense Maori or Napoleon wrasse (*Cheilinus*) of the Indo-Pacific, which may reach 10 feet and several hundred pounds. One massive Napoleon in the Red Sea actually became so tame it would nudge divers looking for a handout. This behemoth, nicknamed "George," once sucked an entire plastic bag of fish from my hand, mouthed it for a few moments, then spat out the plastic bag — empty.

The unique swimming motion of the labrids makes them easy to identify despite great disparities of size and color. These fish swim as if someone had cut the control wires to their tails — they swim along frantically flapping their pectoral fins while their bodies and caudal fins trail uselessly behind. When forced to, they will give a kick with their tails, but then it's back to the old hand-pumps. It's almost as if the wrasse's tail is more like an appendix than an appendage.

Labrids have strong canine teeth mounted prominently in the front of their mouths, and display an aggressiveness out of all proportion to their often modest size. Small wrasses will crowd around any disturbed coral or injured reef-dweller and attack without hesitation. I've seen the Spanish hogfish (*Bodianus rufus*) attack a wounded sea urchin and swim off chomping on a mouthful of spines.

The larger (up to 20 pounds) hogfish *Lachnolaimus*, on the other hand, roots in the sand, expelling clouds of sand through its gill slits. Small crustaceans have no protection against the powerful teeth and jaws of this lavishly plumed predator.

Some juvenile labrids prey on small crustaceans by offering cleaning services. In the Caribbean, young Spanish hogfish and young blueheads (*Thalassoma*) provide this service.

One family of small reef fishes has a reputation for aggressiveness that is in a class by itself. These small terrors are, of course, the damselfishes (*Pomacentridae*). Pomacentrids are diverse, ranging from the tiny jewelfish (*Glyphidodontops hemicyaneus*) of the South Pacific to foot-long brutes that inhabit the cold waters of the Galápagos.

Damselfishes are avid and omnivorous predators, attacking algae, small crustaceans, and other fishes with lightning-like charges. Nor are they above nipping a diver many times their size in pursuit of their legendary territoriality. No other reef creature defends its territory like a pomacentrid, hurrying this way and that patrolling every inch of its declared range. Huge parrotfishes, angelfishes, even rays will swerve out of its way and swim swiftly past the limits of its territory.

Sometimes this aggressiveness is a matter of poignant necessity. Pomacentrids lay their eggs in adhesive clusters on dead coral surfaces. The eggs are easily visible; in many cases they are bright orange, purple, or pink on a nearly-white limestone wall. In one huge depression in the coral bottom in Palau, more than 40 sergeant-majors (*Abudefduf*) had deposited so many purple eggs that the entire eight-foot-across bowl glowed purple in the noonday sun.

Sooner or later something triggers a sudden, all-out attack on the eggs by all the wrasses and butterflyfishes in the area. The formerly bucolic scene becomes a frantic battleground. The sergeant majors race to and fro, desperately chasing away the attackers. The wrasses and chaetodonts swarm against the rocks, devouring the minuscule eggs. Then, after a few minutes, the attack breaks off as suddenly as it began. The attackers disperse without a look back, and for some unknown period of minutes or hours the eggs are safe. The sergeant-majors, as if shaken by the frenzy and the loss, dart nervously back and forth on ceaseless patrol.

One fascinating damselfish, *Dascyllus trimaculatus*, has made a particularly felicitous evolutionary adaptation. It lives, with its close relative, the spectacular Pacific clownfish *Amphiprion*, amid the poison-armed tentacles of immense anemones. So complete is this adaptation that it even uses the same mechanism as the clownfish to survive the tentacles' deadly nematocysts: A mucus layer on its skin surface chemically suppresses the firing action of the anemone's stinging cells.

Two other successful predator fishes, the goatfishes and the parrotfishes, deserve special mention for their very different methods of predation.

The goatfishes (*Mullidae*) are identified by the two long barbels mounted under their chin. These unique appendages operate under quite versatile muscle control, and function almost like tiny hands. The goatfish flutters the barbels beneath the surface of a sandy bottom. From the cloud of sand thus churned, small crustaceans are greedily plucked and eaten. Goatfishes are so successful in uncovering copious quantities of food that freeloaders gather. Wrasses, young jacks, even coneys or angelfishes may join the party. Indeed, I've seen young bar jacks (*Caranx*) tracking so closely above a wandering goatfish that it would have been hard to slip a playing-card between the two of them.

One of the most impressive of all the predation methods on the coral reef is the singular attack of the parrotfishes (*Scaridae*). The parrotfishes are well known to divers, of course, from their size and their namesake beaks. The beaks are really fused teeth, and

massive ones. (Certain of the pufferfishes — Tetraodontids — have somewhat similar beaks, and having seen them in action a diver is well advised to keep his fingers out of their way.)

Parrotfish eat algae, coral polyps, and small molluscs, but they do it in a most unusual, indeed a trademark fashion. Soaring in like torpedoes they take shuddering, crunching bites of solid coralline limestone; that limestone may contain polyps, or be coated with algae on which small molluscs are feeding. It all goes down, as they say, including pulverized chunks of the limestone itself. Within its prodigious digestive tract the scarid separates out the nourishment and, in nature's own boon to vacationers, crushes the limestone and expels it as fine beach sand.

Parrotfishes range in size up to several hundred pounds, and the amount of sand they produce is prodigious. Consider for a fact that a parrotfish eats constantly throughout the daylight hours; all of that ingested limestone comes out constantly, too. I've watched a medium-sized parrotfish taking a chunk out of the reef on average every 20 seconds and expelling sand once every minute. Do a little multiplication (60 minutes × 8 hours × 365 days) and you'll realize that a parrotfish may let fly between 150,000 and 200,000 times a year. You don't need much sand in each incident to pile up a lot of beach.

Some of the larger parrotfishes are very large. In the Caribbean the rainbow parrotfish can weigh fifty pounds, but in the South Pacific there are parrotfishes (*Ypsiscarus orifrons*) that rival the huge Napoleon wrasses (*Cheilinus*) in size. These prodigious parrotfishes have thick, bony bumps on their foreheads and closely resemble browsing buffalo. I've seen a dozen foraging together, leaving billowing clouds of sand so fine that it obscured visibility. In fact, it seems clear that the generally reduced visibility on tropical reefs in the late afternoon may well be the fine sand of uncounted events in nature's parrotfish-relief program.

Not only do the parrotfishes leave the water murky, they also make a lot of noise in doing so. Jacques Cousteau once called the undersea the "Silent World," but nobody told the parrotfish. On a healthy reef free of noisy scuba divers, hold your breath and you'll hear a cacophony of crunches and scrapes as the parrotfishes impact on the coral mass.

Parrotfish identification, like that of wrasses, is a matter of some dispute — juvenile color phases, regular breeding males, females changing sex to become brilliantly decorated supermales all confuse the issue. Moreover, some parrotfishes seem to have the ability to modify their color to fit into their reef background.

In examining this by no means exhaustive roster of swimming browsers I hope I have dispelled the feeling you may have had that what distinguishes fish is perhaps their color and/or size. Nothing could be further from the facts.

When you visit a coral reef, you've entered a community whose species have evolved a host of different, often even complementary, methods of predation. Everything on the coral reef seems to affect something else or depend for its survival on something else. The coral reef community thus presents a complex tapestry of very different events and individuals merging into a harmonious, stable, and effulgent whole. Left undisturbed, the reef community has a powerful drive toward health and growth. Even such parts of the process as may seem destructive (the parrotfish reducing coral to sand, for example) is in the overall scheme a source of ongoing maintenance and renewal.

In the cold waters off New Zealand browses the Mado Atypichthys (New Zealand).

This Spanish hogfish, Bodianus rufus, has attacked and dismembered a sea urchin (Virgin Islands, Caribbean).

The long snout of the butterflyfish Forcipiger longirosthris *protects its eyes as it browses upon stinging corals.*

At certain seasons of the year jellyfish seem to come under fierce attack, here by a pair of Heniochus acuminatus (Truk Lagoon, Micronesia).

*Very common in the Red
Sea and found browsing
across the South Pacific is
Chaetodon lunula (Red Sea).*

This butterflyfish,
Anisochaetodon lineolatus, *is
larger than many angelfish
(Maldive Islands).*

Chaetodon meyeri *is common
in the Indian Ocean, but rare
elsewhere (Maldive Islands).*

The powerful beak of the parrotfish Scarus sordidus *enables it to carve out chunks of coralline limestone (Philippines).*

Passing curiously in deep water comes Chaetodon chrysurus, *with its red-flag tail (Red Sea).*

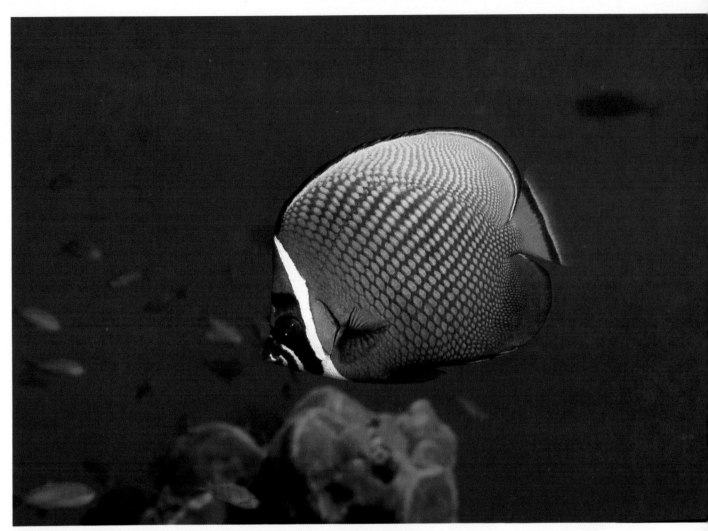

Chaetodon collare *is endemic to the Indian Ocean, where it is ubiquitous on shallow reefs (Maldive Islands).*

Large hogfish (Lachnolaimus maximus) *scoop mouthfuls of sand, straining out small crustaceans and expelling the sand through their gill slits (Grand Cayman, Caribbean).*

This is the juvenile color phase of the intelligent and voracious triggerfish Balistoides conspicullum *(Fiji).*

Here caught between nips at coral polyps, is Tetra-chaetodon plebius *(Coral Sea, Australia).*

Only in the Caribbean can you find the Queen angelfish Holocanthus ciliaris *(Grand Cayman, Caribbean).*

This young wrasse, Plecto-rhynchus chaetodontoides, was thought to be a new species until careful research proved it to be merely a juvenile color phase (Fiji).

Unique to the Red Sea is the lemon butterflyfish Chaetodon semilarvatus (Red Sea).

Constantly busy browsing on coral is Gonochaetodon triangulum *(Fiji).*

Pygoplites diacanthus, *the King angelfish, is widely distributed in the Indo-Pacific and Red Sea (Red Sea).*

Chaetodon falcula *is a fear-less butterflyfish common on many Indo-Pacific reefs (Coral Sea, Australia).*

Rarely seen even in the Phil-ippines and New Guinea is Euxphipops navarchos, *the golden angelfish (Papua New Guinea).*

An angelfish color scheme as intricate as this deserves the name Euxphipops xanthometopon *(Maldive Islands)*.

This small angelfish with the puppy-dog nose is Apolem- ichthys trimaculatus *(Fiji)*.

The French angelfish (Poma-canthus paru) is a fixture on popular Caribbean reefs *(Grand Cayman, Caribbean).*

6

Swimmers in the Open Sea

When we explore the world's shallow coral reefs we are on the borders of a vast and often misunderstood world — the open sea. Divers see some of the denizens of that world, but only when their paths bring them briefly near the scattered reefs they dive. What is impossible for us to truly visualize is their life in that open, endless blue world. We could, without too great a leap of imagination, consider what the life of a reef fish would be like, because the reef itself gives structure to that life. Life follows the contour of the reef, and in that sense is comfortably two-dimensional. It has recognizable boundaries we can see and understand.

But there is nothing in our human experience that would prepare us to imagine the life of an open-water shark or whale or ray. I'm not thinking of the romantic drivel which portrays these creatures as something like big humans off for a swim; I mean the great, special solitude they really experience, when from any quarter may come prey — or predator.

Compared to the familiar coral reef environment, the open seas are as boundless as space. In most seas the ocean floor is miles deep, encompassing millions of cubic miles in volume. These vast spaces are relatively empty, especially compared to the dense life of the reef. On the reef, as we saw in chapter 1, the coral polyps themselves are a colossal food supply which in turn feeds a rich community. In the open seas, however, the base of the food chain consists of the plankton, including pelagic larvae of various species, and the small fish that feed on them.

Some creatures of the open sea feed directly on the plankton and larvae, growing to immense size despite the microscopic nature of the food they eat. Whales, whale sharks, and manta rays are in this category. They swim through space like starships with groping mouths, scooping colossal amounts of nourishment as they pass from night to night.

Of these creatures, we are likely to see and swim with only the manta rays. That's because these immense flying wings look for the concentrated feeding they find at certain points of tidal outflow. In breaks or channels on coral reefs, the mantas may be able to hover and have the currents bring them a rich harvest of food. I've seen them in Palau, for example, in a shallow, narrow, man-made channel near Ngemelis pass. There, in just a few feet of water, rays with 12-foot wing-spans hover with their cavernous maws held wide open.

On some rare occasions, the mantas actually

interact with divers. In Baja California, in Belize (Central America), at Apo Reef in the Philippines, divers and rays have shared moments of magic together. In each of these areas, mantas have shown such curiosity and fearlessness that humans have actually touched or even ridden them.

Like all good things, these encounters can be taken too far, and clumsy humans have frightened off the gentle rays by charging at them. Others commit the unpardonable sin of hunting these docile giants with spear guns. Few punishments suit the enormity of this crime.

Fortunately, most encounters between the mantas and man are benign.

Still, these human encounters offer some rare insights into the lives of these open-water creatures. For one thing, though they live in the void of inner space, they may be as capable of play as puppies. Secondly, though they live in a world portrayed as filled with predators, many reach great sizes and advanced ages; that attests to the relative emptiness of the open sea.

Third, we see that for all their gentleness they are in fact marvelously efficient predators. No slashing charge for them. Indeed, the principal aspect of the open sea for them is its lifelong role as a colossal bowl of soup.

In this planetary bouillabaisse they soar, mouths open, sometimes over mighty spans of ocean. Whale sharks and whales cover long distances, some whales migrating annually from the Arctic or Antarctic to tropical seas and back. While migrations such as that of the snow goose fill us with awe as the sky fills with wings headed south, these whales migrate the same distance but with little fanfare. Only those who follow the whales ever glimpse their solitary voyages, coming up for a breath of air every mile or so, otherwise invisible under the sea's cloak.

The risks to large animals of the open water are formidable. In the cold seas of the world, pinnipeds (seals and sea lions) seek fish in the open water around rocky islands. Yet these teeming waters are also the home of the great white shark. The shark prowls silently, appearing without warning to devour a sea lion. When we swim with the sea lions, their playful nature lulls us into forgetting the fate that awaits them just around the next corner.

I have also heard of a huge manta ray being slaughtered by killer whales (*Orcinus orca*). When we swim with these giant rays, they seem to us lords of the sea. In truth, however, they are but a link in the food chain, and only the vast empty distances of the sea defer those fatal encounters with the ultimate lords of darkness.

Near coral reefs we find schools of fusiliers (*Caesio*), following the reef surface closely. As we move away into open water, we'll find that other species school, too. Great schools of silvery fish form lateral walls of life in countless sites all over the world. Off Palau, in the Red Sea, in the Galápagos, and off Australia, I've seen schools of fish so thick they completely obscured one's vision of the water beyond.

In Australia and other sites countless small anchovies swirl in a predator-confusing ballet, their formation flowing this way and that as if they were a single, huge, fluid organism. This type of schooling is for precisely that purpose, making it difficult for a predator to isolate a single individual to attack. In shallow water, where such schools will sometimes seek refuge, I've watched jacks make passes at the school; the shimmering curtain of anchovy will move gracefully to avoid the charge. Sometimes a jack will charge right at the school; miraculously, a tunnel will form and the frustrated jack will emerge unfed on the other side. Often, this schooling behavior will protect its members flawlessly.

In other cases, though, schooling may not protect the individual. On a reef in the Maldives (which could have been a reef anywhere) I once watched as a school of snappers moved a few feet above the coral. Suddenly a very large jack exploded out of the blue, and faster than the eye could see this silver streak plucked a snapper from the school.

In the blue waters off Ras Muhammad in the Red Sea, schools of snapper numbering thousands of individuals will form a colossal cylinder of fish, slowly revolving like some living funnel cloud in the open space just out of sight of the reef.

It need not even be prey who school, though a school in their case may represent the tenuous security of community. Predators also form great congregations in open water — for breeding perhaps, but also, it appears, for socializing.

In certain regions I've seen dozens to hundreds of barracuda gather, form a coherently moving school, and soar together. They were not feeding; there seemed, in fact, to be no interaction other than schooling for company. When I swam out into the middle of one such school in Palau, I was enfolded in an immense revolving cylinder of barracuda so thick I could not see out through their bodies.

It was as if these placid though fearsome-looking predators were on holiday, gathering together just for the anonymous companionship of a crowd. Since barracuda are known to practice solitary hunting, their social schooling is even more noteworthy. For a solitary human, it is also a sight to fill the mind with wonder.

Other predator species school constantly. Killer whales hunt in pods, sharks will often be found in loose aggregations in particular sites. The amberjacks, sleek sliver wolves of the sea, flow from the open sea to the reef and back, ever on the prowl.

In the open blue of the sea, fascinating relationships occur. Divers who regularly leap into the blue ocean miles from shore will find porpoises, pilot whales, and — under certain circumstances — even humpbacked whales moving along roughly parallel to shore.

And yet, it seems that behind each such school follows one or more shadowy of the predators like the oceanic white-tip shark, *Carcharhinus longimanus.* The sharks may follow to pick off stragglers, to try to snatch the young, or for some other reason unknowable at least for now.

For a solitary human, the blue sea is most enormous when you are alone in it with one of these open-water hunters. What strikes you first is that there is no place to hide. You may even spend a moment envying the flying fish their special skills; if only you, too, could leap out of the water and soar a hundred yards. . . .

But there is no such escape, and these cold-eyed predators seem to know that at a glance. Circling warily, they watch every move you make. There is an inevitability about their stalking, as if they, too, know they need not hurry. Their prey has nowhere to hide, so they need to take no risks.

I've tried to evoke the nightmare of being alone in the blue sea with a predator, because that is how so many of the pelagic schooling fish inevitably end their lives. Though each of them is a predator, there is always a larger predator in their future. For tiny prey the end may be a flying fish or houndfish; for the flying fish it may be a barracuda or wahoo; and even the largest predators must inevitably face the shark.

Nature has provided some defenses, of course — the flying fishes' long flight, the blue or silvery body color as a camouflage, and the advantage of schooling; but in the end the steel-eyed predator waits.

On some occasions I've watched frightened small parrotfish press tightly together in a large school, paced by sharks. I was struck by the fact that these panicky fish swam so close together and so much in unison that they resembled a single large animal. I wondered whether this was mere coincidence, or more likely part of the incredibly balanced design of nature. What better way to avoid attracting some large predator watching from the distance than to impress it as being large, too?

The really swift open-water fish need not resort to such behavior, since they can out-run most predators. Schools of rainbow runners, amberjack, or little tuna are loose aggregations of lean, silvery speedsters. Swooping in from the darkness, they briefly eye the human oddity, then soar off into anonymous darkness again.

It is important not to impulsively categorize these fish by a single behavior pattern, because they may actually have a much more varied repertoire. For example, horse-eye jacks (*Caranx latus*) will sometimes swim in groups of four or five individuals, especially when near a sheltering reef. At other times they will form loose, long streamers of hundreds of slow-moving individuals. At another site, several dozen may mill about above shallow coral heads in only twenty feet of water. During these encounters I've sensed a great ease, a carefree schoolyard feeling, almost as if like puppies these fish had endless time ahead. Sometimes members of the school will stream off and swim over to investigate the human intruder. Yet there is always a natural edge of caution, a bright-eyed wariness — the brittle edge of the intended victim.

There is an undersea arch in Fiji where up to 300 of these silvery jacks can be found all looking out the archway like the crowd in a football stadium, all eyes. A mild current flows through the arch, so the jacks attain not merely security, but a flow of oxygenated water as well.

This is the other side of the predator's life — long, lazy days of schooling, streaming along with your friends in search of food. This lazy life may last for months or years, until the day of your meeting with your own predator.

That makes life in the open water an interplay of light and darkness, joy and sadness, carefree play and sudden death. It can be imagined only at its margins, for we humans can never experience it fully and return to tell the tale.

Near the Yongala wreck in Australia, my diving partner warns me off as she maneuvers close to a large sting ray (Great Barrier Reef, Australia).

Predatory hawks gather for
the inevitable denouement as
a frantic mother sea lion tries
to revive her stillborn pup
(Galápagos).

These blonde sea lions are
the principal prey of the great
white shark, Carcharodon
carcharias (South Australia)

Sea lions (Zalophus
Californicus galapagensis) *are
inordinately curious, and like
children test everything by
biting (Galapagos).*

It is dusk, and this hawksbill turtle (Eretmochelys imbricata) has settled for the night (Roatan, Caribbean). ▶

In open waters, schools of fish numbering thousands will form themselves into virtual walls (Galápagos).

Thinking me another turtle as I held my breath, this green sea turtle was startled when I released the camera shutter and exhaled (Papua New Guinea).

The blue-spotted ray, Taeniura limma, is a gentle creature which arouses our curiosity with its gay color scheme. Why the spots? (Red Sea).

*Graceful butterfly of the open
blue water is the manta ray
(Philippines).*

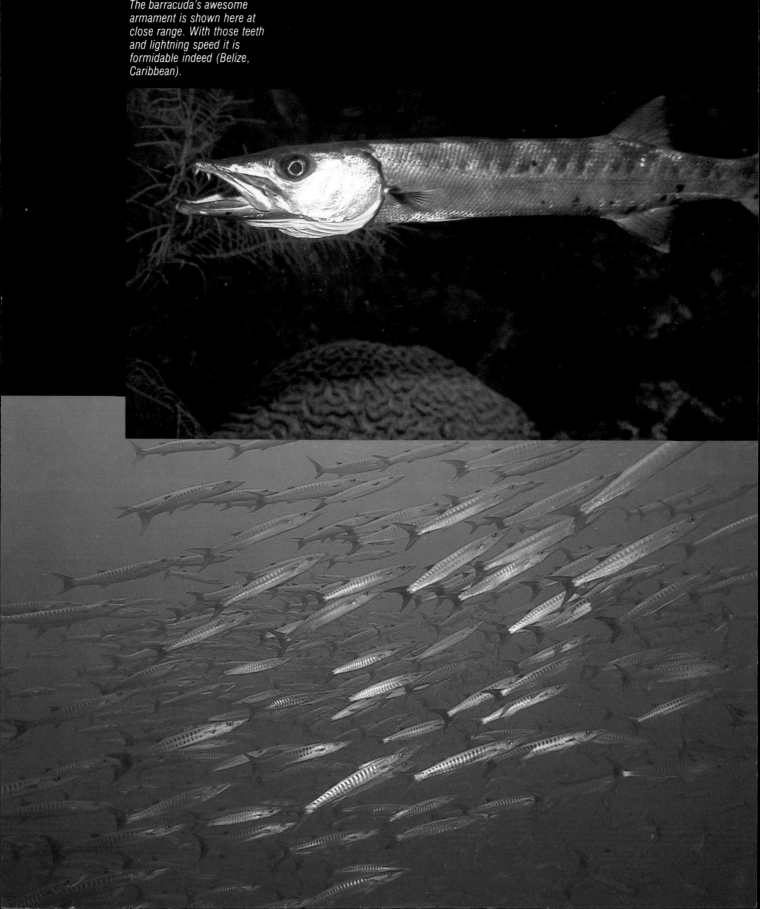

The barracuda's awesome armament is shown here at close range. With those teeth and lightning speed it is formidable indeed (Belize, Caribbean).

Schooling anchovy lose their individuality and become part of a silver curtain, confusing potential predators (Coral Sea, Australia).

7

Odd Couples

One of the most intriguing sights on any coral reef is the social interaction between two different animals. One major interaction, of course, is for one creature to be eaten by another; but there are also other relationships of dizzying complexity, of surpassing strangeness and beauty.

There are several types of interaction in which two animals permanently share living quarters, or in which one animal even lives out its life upon the body of the other. Depending upon the degree and quality of interaction, these relationships are termed mutual, commensal, or parasitic. Briefly defined, mutualism (often referred to as symbiosis) produces benefits for both parties; commensalism benefits only one party but does not harm the other; and parasitism (which will be treated more fully later) benefits one party at the expense of the other.

Among the mutual beneficiaries are some of the sea's most photographed and most beautiful citizens. The anemones, or flower corals, host several different symbionts. In the Caribbean, anemones often have small shrimp of the genus *Periclemenes* dancing amid their nematocyst-armed tentacles. Even when a wave-surge lashes the anemone's lethal arms, these shrimp dance nimbly and invulnerably among them.

In the case of *Periclemenes yucatanicus*, we may see a clear benefit to the shrimp in the relationship. After all, without the surrounding tentacles, the shrimp would be eaten in a moment. What, however, is the benefit to the anemone? It is possible that the crustacean cleans sand or other debris from the tentacles; it may conceivably even lure would-be predators to entrapment in the anemone's grasp. This is really speculation, however, as the full relationship is simply not understood as yet.

As we shall see in several examples, similar rituals and relationships exist in both the Caribbean and the Indo-Pacific, but with totally different participants. In the Pacific, the principal commensals found amid the anemones' tentacles are the clownfishes, *Amphiprion* and *Premnas*. These clownfishes are among the most brilliantly colored of all tropical reef fish, and their repertoire of behavior seems more varied than that of the Caribbean shrimp commensals. They will, for example, swim several feet away from their host to snatch plankton from the food-stream (or in some cases to nip an intruding diver).

When the host anemone has snared a meal, it often inverts its tentacles so that they disappear inside the balloon of its body. At these times, the clownfish is completely dispossessed, and at some risk. If a predator happens by, the clownfish may not survive. This seems to be a complex game of chance, for I have seen some clownfish survive for several years in the same anemone, while others have disappeared from the host within a single season. I rush to point out that my observations were occasional, and do not constitute a scientific study; therefore any conclusion is purely my personal impression.

Still, even these sporadic observations suggest a pattern of protection that is at best incomplete. Anemones survive for spans of ninety years and more; their commensals come and go. This would further suggest that while the anemone can do quite well without its symbiont, the shrimps and clownfishes may be quickly eaten if deprived of the protection afforded by their lethal host; thus the mutualistic relationship has the greatest value to the more edible partner.

One further dimension of this dance of mismatched partners is the behavior of the damselfish *Dascyllus arenatus*. This small black damselfish with stark white spots often dances about the same anemone claimed by a clownfish. For the underwater photographer, taking a picture of a clownfish can be doubly challenging when several tiny damselfishes are milling about in front of the camera as well.

The mechanism these fishes use to survive their host's nematocysts is nothing short of amazing. The anemone becomes accustomed to its symbiont by recognizing at the molecular level a mucus on the fish's skin. Experiments in which the mucus was wiped off the clownfish's body resulted in the host stinging the now-unrecognizable clownfish to death.

That may be why clownfishes are constantly touching or rubbing their anemones' tentacles. Talk about the risks of being out of touch. . . .

Though this discussion of anemone symbionts has focused on the defensive aspects of the relationships, we must not forget that each of the individuals is itself a predator. Clownfish, damselfish, and shrimp browse on organic debris, plankton, and larvae. All prey in order to live.

That fact leads us to an introductory facet of the sea's most compelling social phenomenon. This major element of social behavior in the sea is the ubiquitous cleaning ritual. The cleaning phenomenon arose as the necessary counterforce to that damaging condition called parasitism. Many fishes are infested with surface parasites, usually small crustaceans which skitter about

on the host's skin or gills. Scientific studies which removed all cleaner animals from a sample reef showed that, unchecked, the parasites eventually overwhelm each fish host and cause its death.

Nature's elegant solution is to have other small animals prey upon these parasites. The elegance, of course, is that the drive of predation motivates the cleaners to perform a vital service.

There is a wrinkle, though. Most of the cleaners are natural prey of the hosts. Groupers, for example, just love to eat shrimp and small fish. How can these wary morsels be convinced to leap aboard the grouper's body, indeed *enter its mouth*, in pursuit of the parasites? Only by a firm truce, communicated by an elaborate series of unmistakable signals.

After all, why should the small shrimp and fish who act as cleaners take on any risk? They are perfectly capable of finding other prey.

Out of all this evolved the *cleaning ritual*. The ritual aspect is the unvarying repetition of a set of signals clearly understood by both sides. The cleaning ritual resembles human behavior such as ringing a doorbell, or putting coins in a slot to gain access to a bus. The trigger action is unmistakable and invokes an entire series of conventions which are socially understood to follow.

Interestingly, while the trigger action itself can vary from species to species, it is identical within any species. Some species such as bonnetmouths, small wrasses, and parrotfishes will approach a cleaner and assume a vertical position. Since it would seem that standing motionless on its head or tail is both difficult and tinged with predation risk for these small fish, this signal appears clear — the host makes itself vulnerable to declare its peaceful intentions.

Other fish, such as goatfishes, will approach a cleaner, become motionless, and even change body color, thus rendering the parasites easier to see. Large animals such as moray eels or groupers will glide up to the cleaners and become motionless. If they have parasites they will even hold their mouth wide open.

On the other side of this cleaning equation are the cleaners. These may be shrimp of several species, small gobies, small wrasses of particular species, and the juveniles of specific angelfishes and hogfishes. One extraordinary aspect of these cleaners is the prevalence of a pattern of stripes in their body colors. Some of the shrimps and all of the gobies and wrasses are boldly striped; moreover, the cleaning angelfish juveniles are striped *only* during that portion of their youth when they provide cleaning services.

One marvelous irony of this complex system is that the cleaner shrimps are actually preying upon isopods and copepods which are closely related to them.

The cleaning ritual seems to have three distinct stages which we could call the turn-on, the service, and the turn-off.

I've already mentioned the turn-on. If you are snorkeling or diving on a reef and see one or several fish laboriously holding themselves in a vertical position, you are observing a cleaner station. In the case of a larger animal, such as a moray eel or grouper, you'll spot the host lying motionless, often under a shadowy ledge; in some cases you'll see its mouth held open, which is a sure sign that cleaning is taking place. In these cases, try to insinuate yourself as close as you can without disturbing the process. You may even get close enough to actually see the tiny parasites moving about on the host's skin.

There are some large, unsightly parasites. In the case of certain fish and in certain parts of the world, you can even predict where the parasite will be. For example, in Bonaire, most creolefish (*Paranthias*) will have a one-inch-long isopod on their cheeks under one or both eyes. In the Cayman Islands, the same parasite will be on the cheeks of all the barred soldierfish, *Myripristis*. Yet in the Turks and Caicos Islands, the host is the four-eyed butterflyfish, *Chaetodon ocellatus*.

Watching the cleaners at work is fascinating beyond description. When a huge grouper opens its gills, you can literally see all the way down to its gullet, and right out through the open gill slits. There will be a tiny wrasse darting about inside, completely ignoring the cavern-like maw and the huge teeth, and looking for all the world like a maid cleaning up a hotel room.

Cleaning may last for a few seconds to several minutes, and during that time both cleaner and host are in an exposed and vulnerable condition. What happens if some large predator (or a human who certainly looks like a large predator) approaches?

As you would expect, such a formalized system has built-in escape mechanisms. Smaller host fish, e.g., wrasses, parrotfishes, or bonnetmouths, give a little shudder, then rapidly swim away. Interestingly, while the cleaners jump off, they do not seem to rush for cover. It is as if these cleaners know they are still under the truce until they reach their station; this is particularly true of the cleaner juveniles, gobies and wrasses. The shrimp are more wary, never getting very

far from shelter. Apparently, these crustaceans are just too tasty a morsel to afford any carelessness.

Larger host animals seem very reluctant to break off their sessions of being cleaned. Perhaps they feel less vulnerable, or perhaps they suffer greater stress from the parasites. In any event, these larger hosts will often stay in place until the very last moment, allowing divers to observe the cleaning ritual at very close quarters.

One final marvel of the cleaning ritual is a delightful fillip on this complete process. It is as if nature, in allowing every possible niche to be filled, almost exercised a sense of humor. The Pacific cleaner wrasse, *Labroides*, is closely resembled by a mimic known as *Taeniopterus*. This false cleaner eats fish scales.

Imagine then, the parasite-plagued host which sees up ahead a striped fish closely resembling *Labroides*, doing the same characteristic bobbing dance as the cleaner. Looking for relief, the host approaches to be cleaned. It flares its fins, opens its mouth — and gets a ferocious nip from the false cleaner. I've seen furious hosts chase the false cleaner after such a nip, but by then it is, of course, too late. In other instances, I have seen clear cases in which the intended victim recognizes the false cleaner and chases it away before it can bite.

In one famous controlled experiment, a false cleaner was introduced into an aquarium with a grizzled old veteran grouper. The grouper watched as the fake did its convincing dance.

When the false cleaner came close, however, the grouper simply snapped open its gigantic mouth and gobbled up the hapless impostor. Predation is, indeed, a two-edged sword.

For careful divers, the many variations on the cleaning ritual offer one of the most rewarding subjects we can observe on the reef. While patience and approach technique are crucial, the sense of communion and understanding we take away from these observations makes the effort most worthwhile.

There are other predation-based relationships which involve oddly matched partners. One of the most unusual of these is the commensal pairing of the blind prawn and the seeing-eye goby. This Felix and Oscar pair may be found on shallow sandy bottoms across the Indo-Pacific.

The first sign the cautious observer will spot is a motionless goby two to three inches in length sitting beside a small burrow-opening in the sand. The goby is white, with orange-brown bands girdling its body.

If you sit very still, you will soon see a modest disturbance within the entrance of the burrow. After a moment, like a tiny bulldozer at work, a gray prawn will push a mound of sandy debris out of the hole. Time after time this performance is repeated, as the prawn, like some compulsive maid, maintains the burrow. During all of this exertion by the prawn the goby simply sits, its tail normally pointed toward the burrow.

If you approach a bit closer, the goby will move closer and closer to the burrow, eventually placing its tail across the burrow's mouth. The shrimp, encountering the goby's tail, is warned of possible danger. A closer approach will cause both the goby and the now thoroughly alarmed prawn to dive down the hole for shelter.

Interestingly, while their defensive system is shared, the goby and prawn feed differently. The prawn eats small crustaceans and organic debris found in the sand as it bulldozes. The goby, a surface hunter, dines on passing plankton borne by the current as well as small crustaceans scuttling across the sand.

Another wonderful symbiosis is that between the remora family, the Echeneids, and various large, free-swimming hosts. The remoras and sharksuckers are specifically adapted to open-water commensalism. So specialized are they morphologically that people who know few of the sea's creatures instantly recognize these.

Atop each remora's head is a large, ovoid rim bordering a retractable plunger. When this fish swims up next to a smooth surface of any kind (even the hull of a ship!) it makes rim contact, uses specialized muscles to pull the plunger, and creates a vacuum cup. Once attached, the remoras may remain attached for hours, days, or indefinitely. I've spoken with divers who attempted to dislodge large remoras from manta rays and found they could not do so short of injuring the animals.

Remoras may grow to a length of nearly three feet, and are occasionally found swimming alone, apparently seeking a new host. Small ones tickle when they try to attach to your exposed skin; they may also attach to your rubber suit or any other surface.

These echeneids ride sharks and rays, expending little or no effort in the endless search for food. They eat scraps of food dropped by their host animals.

Another type of passenger for sharks and rays is the small, vertically striped pilotfish, *Naucrates ductor*. This is remarkably skillful at riding the bow wave of the larger animal, which reduces its own swimming effort. Both remoras and pilotfishes escape predation by thus allying themselves with their large hosts, besides reducing greatly their own exertion in roaming the open water in search of food.

It is not at all clear that the ray or shark hosts achieve any benefit from these commensal hitch hikers. Almost certainly the rays, as plankton-eaters, have no need of their commensals; it is almost comical to think that a cruising shark derives any important benefit from the remora riding on its back.

While the pairings I have mentioned are some of the most characteristic of the phenomenon, there are many others.

Some tiny fish are commensal with free-floating jellyfish, protected from predation by the proximity of the jellyfish's stinging tentacles. Pearlfishes (*Carapus*) live out their lives in the anal cavities of sea cucumbers. There are a number of species who live within corals and sponges; look inside almost any large Caribbean tube sponge and you'll discover neon gobies fluttering in the sponge's excurrent wash.

There are even invertebrate species (small crustaceans, principally) which are completely embedded within the sponges' tissue for their entire lives. Certain other crabs live in coral crevices until the growing coral entombs them (except for a small opening through which they feed). Another spectrum of fascinating creatures live within the tangle of feeding arms spread by crinoids. Small galatheid crabs, shrimp, isopods, copepods, brittle starfish, and Lepadichthys (the only fish in the sea with a neck), may all be found together on a single specimen of crinoid. In this case, the crinoid not only provides protection; since it is constantly feeding with its arms, these commensals can take a percentage without harming the crinoid.

So specially adapted are these commensals to their individual crinoid that they are polychromatic. Their body color matches or complements that of the specific crinoid on which they live. Considering that crinoids are found in all shades of brown, yellow, red, green, orange, and blue, this is indeed a brilliant assemblage.

These foregoing relationships share a characteristic permanence, many pairings lasting for the lifetime of the partners. At a more common level, there exist some pairings of brief convenience.

Often on a sandy bottom, one will see a stingray buried up to its eyes in sand. When it is disturbed, it will rouse itself and swim away, streaming contrails of sand. Almost immediately, a juvenile jack will appear from nowhere and begin riding close to the ray's back. This pairing appears to offer the jack some food from the small crustaceans stirred by the feeding ray. It does not seem that there is any benefit to the ray in the relationship.

Similarly goatfishes, browsing on the sandy bottom may be accompanied by small wrasses and groupers;

the goatfish stirs up the sand with its chin-mounted barbels, and its companions pilfer some free food.

In the Indo-Pacific, large triggerfish such as *Balistoides viridens* feed by overturning small coral chunks; they may be surrounded by a motley crew of other reef fish all feeding off the churned-up organic material thus released.

What is fascinating here is the different biological base of this behavior. In the case of the polychromatic

Pairs of eye-catching Bantayan butterflyfish (Chaetodon adiergastos) *hover amid the Philippine reefs (Philippines).*

crinoid commensals or the blind prawn with its seeing-eye goby, evolution paired these creatures eons ago. In the case of the jacks and rays, or the followers of goatfishes and triggerfishes, it almost appears to be learned behavior without necessarily a long evolutionary history. We might think of these pairings then as falling into two broad categories — *evolutionary* and *opportunistic.* Evolutionary pairings are forever, as it were, while opportunistic pairings last only for the moment of benefit.

As you can imagine by the examples cited in this chapter, the more skilled you become as a diver/observer, the more of these unusual pairings you will see. In a way, it's like birdwatchers cataloging species they spot on their forays. There is such a complex tapestry of pairings on any active reef, however, that you may spend years gathering your list of odd couples.

Butterflyfishes pair for life, and few creatures seem more forlorn than a butterflyfish which has lost its mate (Maldive Islands).

Here a seeing-eye goby
(Cryptocentrus) *places its tail
across the burrow to warn the
blind prawn* (Alpheus) *not to
emerge (Red Sea).*

Finger sponges can be nearly
smothered with symbiotic
zoanthids (Cayman Islands,
Caribbean).

The clingfish Lepadichthys
*lives amid the complex arms
of the crinoid* Comanthina
(Coral Sea, Australia).

Sometimes a cleaning job attracts not one but several gobies (Belize, Caribbean).

A tiny cleaner wrasse swims around within the gaping mouth of a coney, Cephalopholis miniatus (Maldive Islands).

The unusual sucker disc of the remora has a muscular central plunger which can be lowered to pull a vacuum (Ponape, Micronesia).

A grouper, Cephalopholis miniatus, *and a pufferfish,* Arothron hispidus, *lie down side by side to be cleaned. A shrimp is on the grouper's lip while a wrasse tends to the puffer (Maldive Islands).*

The scar on this sweetlip is from a large parasite which had probably spent years attached to the same spot (Coral Sea, Australia).

*In the Indo-Pacific the wrasse
Labroides dimidiatus replaces
the goby as a cleaner fish
(Maldive Islands).*

The intricate symbiosis between clownfish (here, Amphiprion perideraion) and their host anemone is one of the sea's most famous pairings (Philippines).

Almost every manta ray and shark we see has one or more remoras (Echineus naucrates) attached (Philippines).

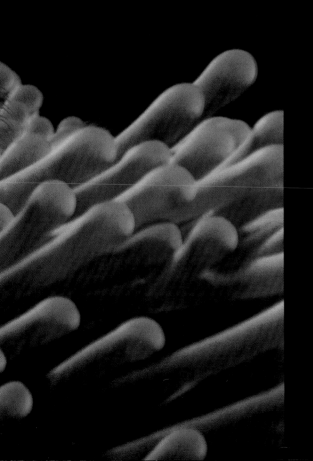

Once truce is declared,
gobies pick parasites from
the body of a green moray
Gymnothorax funebris
(Belize, Caribbean).

This colorful puffer tries to ignore the photographer in order to complete its cleaning session (Philippines).

A large green moray eyes a group of cleaner shrimp as it approaches for service (Galápagos).

The clownfish Amphiprion bicinctus *nestles among the tentacles of a colorful anemone (Coral Sea, Australia).*
◀

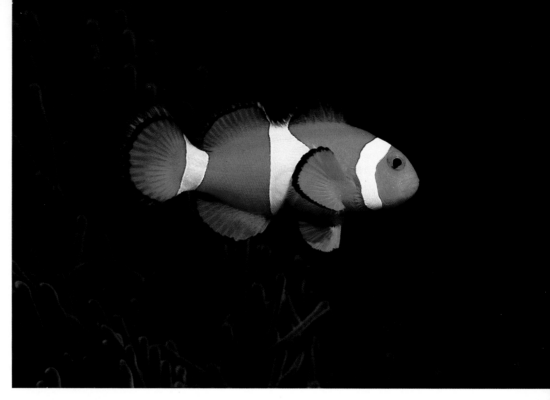

The dazzling clownfish shown here is Amphiprion percula *(Philippines).*

Butterflyfish such as Forcipiger longirostris *mate for life, and dine daintily on coral polyps together (Fiji).*

8
Masters of Disguise

In our earlier chapters we described things divers can see for themselves on any active coral reef.

However, while the divers are observing these visible animals or relationships, they will be under careful scrutiny by some masters of disguise they probably won't see. These creatures use a variety of methods, but have one thing in common. It is possible to look right at them without realizing that they are there.

By far the most resourceful and impressive of these masters of disguise are the cephalopods, which include the octopus, cuttlefish, and squid.

All of the cephalopods are relatively intelligent, and are possessed of a sophisticated control over their body coloration. The basic color-control mechanism consists of tiny sacs of pigment, each a different color, embedded just beneath the animal's skin-surface. These amazing neurocontrolled pigment sacs are called chromatophores. The pigment in these sacs may be brown, white, yellow, or even iridescent blue. Each sac, when relaxed, is a mere pinpoint, its color almost invisible. However, the cephalopod can, by neural control, stretch or flatten out all of the yellow discs, for example, and suddenly its body color appears as yellow as the rock on which it sits. Moments later the creature may glide into shadow: its white sacs contract, its dark brown sacs expand, and it is suddenly as dark as its new surroundings.

Now, given only that observation, you might conclude that the cephalopod's body-color system responds automatically to changes in its background environment. In fact, the system is far more sophisticated than that. The pigment sacs, which are under direct neural control, are thus a facet of the animal's nervous (hence emotional) system.

That is the marvel of encountering a cephalopod. Like an expressive human face, its body acts as a display for its emotions of the moment — the body becomes palette of emotional colors, as it were.

Recently, on a dive in the Philippines, I came upon a lone octopus under the shadow of an immense rock overhang. Since I had been moving about slowly, photographing various small corals, nudibranchs, and other creatures, the octopus had perhaps already concluded that I was clumsy but harmless. Perhaps the brilliant flash of my strobe even aroused its curiosity.

In any event, when my wanderings brought me face to face with the octopus, it displayed no sign of fear. Remaining next to a particular rock, it sat and watched me. With great caution, I moved in close and took a picture. At the brilliant flash of the strobe, the octopus twitched violently — but stayed where it was. A moment later I took a second picture, then a third.

As I watched the octopus it began to raise bumps all over its skin (another peculiar cephalopod talent) and pulses of alternating brown and white swept over its body. It was like watching chocolate swirl ice cream being mixed; color flowed all over the animal in ways the most sophisticated video game would have envied. Slowly, it began to writhe its way with amazing fluidity across the rubble bottom. As we moved together, I took pictures, and soon I saw that the octopus was returning to its home lair.

This apparent residence of the octopus was a crevice under a moderate-sized coral head lying on the sand. The octopus entered the lair, backing in with its eyes still on me and its body still pulsing with color.

Then, in a marvelous display of octopodal skill, it used its sucker-lined arms to move a series of small coral chunks and build a door between us. Soon I could see but a small hole in the pile of rubble with a bright eye behind it. It is hard to describe the versatility with which this animal uses its body. When you first look at the octopus, the skin between the arms looks like a mere fold of skin; moments later, the octopus may be using this skin like a hand and fingers. The only analogy I can image in humans would be using the back of your knee to pick up pebbles.

I decided to try an experiment. Slowly, piece by piece, I removed all of the coral debris, exposing the octopus once more.

It flushed completely white, began churning its body, and stared at me, seething. I waited for its reaction, and was astonished when it came.

The octopus formed its mantle into a pulsing scoop. I could see it gathering something under its body. Then it moved forward and explosively fired a cloud of sand at me. It was an unmistakable sign of annoyance, and I was so taken aback I neglected to fire the camera.

I waited patiently, letting the creature know that I intended no harm. Sometimes simply remaining very still is a convicing sign of non-aggression on the reef.

The octopus seemed to turn quizzical, and suddenly its body was chocolate-brown on its right side and white on its left, with the line of demarcation running vertically down between its eyes. I almost felt it was a signal in a language I could not read. Feeling distinctly illiterate, I remained very still.

Emboldened, the octopus soon emerged from its rocky shelter and perched atop the rocks in plain view, its tentacles trailing behind.

It was as if my peaceful signals had been received, acknowledged, and were now the basis of an agreement between us. I felt wonderful, in the way of an animal lover with any affectionate animal, only much more so. For the next several minutes, until I ran out of film, the octopus and I shared the opportunity to unhurriedly study each other at close range.

I've recounted this incident at some length to make a specific point: Encounters with cephalopods are richer in content than those with other marine creatures. This is because the human observer gains the impression of dealing with a keen (though very different) intelligence.

Oft-told tales of octopus tricks are legion. Octopus crawling around on shore at night, octopus building shelters, and the famous one about the octopus thief. In a marine lab, it was discovered that an octopus was taking the lid off its aquarium each night, foraging in other tanks, returning to its own tank, *and replacing the lids.* Until the octopus was actually caught in the act, no one suspected it was capable of such maneuvers.

One can only imagine what took place in the mind of the octopus as it replaced the tank-lids to erase all traces of its foray: Smugness? A sense of orderliness? Some other emotion we can't even conceptualize?

Similar to the octopus but somewhat less versatile is its cousin the cuttlefish. As with the octopus, encounters with cuttlefish are intellectually satisfying. The only other creatures with whom I've had that feeling are porpoises, certain groupers, and a few triggerfishes and large wrasses.

Unlike the octopus, whose fluid body has no skeleton but merely a series of cartilaginous rods to anchor its musculature, the cuttlefish has a rigid structure. The cuttlefish's internal frame, or cuttlebone, is a graceful, very light honeycomb structure; the cuttlefish's body has a natural hydrodynamic grace, but with certain embellishments. A muscular fringe about the lateral edges of its body gives it positional control like a tiny wing. Its head with its 10 tentacles can move on what looks like a neck at various angles to the body.

Both octopus and cuttlefish seem to remain in the same reef area for years if they survive predation. Octopus, for all their intelligence, are no match for large groupers or morays. Similarly, cuttlefish may be devoured by sharks, jacks, or tuna in a moment.

It appears, then, that the color manipulation of cephalopods has both offensive and defensive aspects. Indeed, the uses of these skills may even vary between the octopus and its free-swimming relatives the cuttlefish and squid. Octopuses, as bottom-dwellers and mollusc-eaters, probably don't use their disguise offensively. Cuttlefish and squid, however, can hover motionless in open water, camouflage themselves with neutral colors, and snare passing fish who pass too close. The most common tropical sightings of the squid *Loligo* is in small groups arrayed in a line. Sometimes this graceful chorus line will array itself in the shadow of a dive boat, moving in unison to the comings and goings of divers.

Both octopus and cuttlefish have visible siphons or jets through which they can expel water. The octopus, when alarmed, may fire its jets at the intruder, or use them to flee. The cuttlefish has an external siphon protruding beneath its tentacles. With great precision, the animal directs the hoselike siphon in any direction to propel itself. When a cephalopod decides to flee, it aligns the siphon forward and backs away from danger at high speed. The bodies of squid and cuttlefish are particularly streamlined for traveling in reverse.

This unique design suggests for both squid and cuttlefish a slow and cautious offensive style and a flight defense. Their pattern is to glide close to their target, then shoot their tentacles forward to grasp the prey.

Tropical squid tend to be rather small, reaching perhaps 10 to 12 inches in length. Their bodies are cigar-shaped, with large eyes mounted just behind the base of their tentacles. Hovering in open water, they may adopt a complete silver coloration that renders them nearly invisible against the open-water background. In shadow or at night they may take on colors ranging from dark brown to deep red.

Tropical cuttlefish can be substantially larger than squid, with an overall body length approaching two feet. Despite this relative bulk, cuttlefish have an uncanny ability to blend into a reef scene. On some occasions I have discovered cuttlefish disguised as coral heads watching from a few feet away. The way to spot them is to get very low to the coral, and scan for anything that is hovering. Since the cuttlefish avoids having the coral touch its sensitive skin, it will always leave a few inches clearance; often that may provide the only visible sign of a cuttlefish.

While the cephalopods are the most spectacular of the masters of disguise, a number of other reef-dwellers have evolved similar mechanisms.

The peacock flounder (*Bothus*) has chromatophores similar to those of the cephalopods; this flatfish lies on — and blends into — almost any type of sea floor. Flat white sand, coarse volcanic sand, algae-covered rubble — it's all the same to the flounder. In a moment, it can shift its colors from tan to chocolate to blue-ringed confusion, confounding the eye. I've even read of experiments in which these flounder, placed on a black and white checkerboard, adapted their body patterns to blend in.

Sharks, whom we'll treat in their own chapter, are masters of disguise without changing color. These predators have light bellies and dark gray or blue-gray upper bodies, in a pattern known as obliterative countershading.

This method of disguise is elegantly simple, and lethally effective. It works by using natural sunlight shining down from above. As the shark cruises, its darker upper body is lightened by the impinging sunlight, while its lighter belly is in shadow. The result — a gray ghost blending into the serene emptiness of the open ocean, a spirit you never see until it has arrived.

A different use of disguise is used by groupers or sea bass. Many of these serranids have a distinct ability to change body color and pattern, though their system is not as sophisticated or flexible as that of the cephalopods or flounders.

In some groupers the color change may merely be from dark (for hiding in darkness) to light (for blending into an open sand or coral rubble area). Others have combinations of rather greater virtuosity. The Caribbean coney, *Cephalopholis fulva,* for example, exhibits a normal chocolate brown color speckled with blue-rimmed black spots. When the coney is disturbed or alarmed, the lower half of its body fades to stark white, and its ocellated spots contract to mere pinpoints. The coney's preferred lair is a coral crevice with a white sand bottom; hence these colors may serve to blend its upper half with the coral's shadow, while its lower half matches the sand.

One interesting color variation in the coneys is found in certain individuals whose body color is brilliant red rather than brown. In deep water, red tends to look black, and this small (9-12 inches) predator is almost impossible to see under shadowed, deep ledges.

There are many other masters of disguise whose talents are used for defense rather than offense; in our next chapter we'll discuss those animals and their techniques. In many cases, offense and defense practically merge, for an animal that is difficult to see is equally adept at pursuing or avoiding predation. In addition, we'll see in chapter 10 (on ambush predators) that many ambushers also use color and disguise to lull their prey into fatal carelessness.

Before leaving predators whose principal use of disguise is offensive, I should mention once again the swimmers of the open sea (chapter 6). Not only sharks, but also rays and whales use obliterative countershading to escape detection.

In another variation on that theme, consider barracuda, tuna, mackerel, jacks, tarpon, anchovy, and a host of other open-water species which are reflective silver in color. Seen at a distance in open water these silver fish become indistinct, and can easily pass without detection in the miles of open sea. Conversely, all of these camouflaged denizens of the open water can approach their prey undetected using the same coloration.

Now add to their color the fact that most of these creatures are elongate, even projectilelike in body shape. When they approach, their small body cross-section plus their blend of color can remove the critical safety margin small prey try to maintain.

In summary, we may conclude that disguise may range from the virtuosity of the cephalopods to the plain-wrapper coloration of the open-water predators. All are effective, honed by the passage of thousands of centuries of trial and error. Those in error did not survive the trial, so it is really not surprising that today's disguises are highly effective.

They have certainly stood the test of time.

Poised above the coral, the cuttlefish shudders, extends its tentacles, and nestles a tiny white egg amid coral fingers (Philippines).

A pair of cuttlefish nestle together in a coral depression (Philippines).

Not only can the octopus flash constantly changing colors; it also alters its skin texture at will (Philippines).

Here an octopus is being stroked by the tentacle of a second octopus (Fiji).

A disturbed octopus shows this pattern because I disassembled its hiding-place (Philippines).

A peacock flounder (Bothus lunatus) shows its spots as it moves, but it will soon make the color disappear so it can blend into the white sand (Grand Cayman, Caribbean).

Using its lateral frills to steer, the cuttlefish jets away (Philippines).

When the octopus wishes to fade into darkness, it can do so in mere seconds.

*Bold as brass, this octopus
stands out in broad daylight
(Coral Sea, Australia).*

We are constantly amazed by
the octopus repertoire of
disguises.

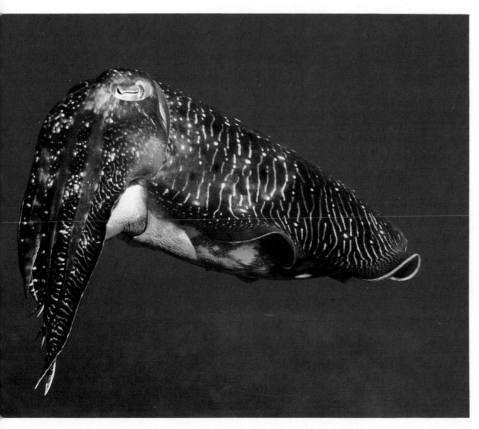

Changing color, the cuttlefish
hovers in open water
(Philippines).

Even the large eyes of the
crocodilefish Cociella are
disguised with decorative
tissue (Philippines).   ▶

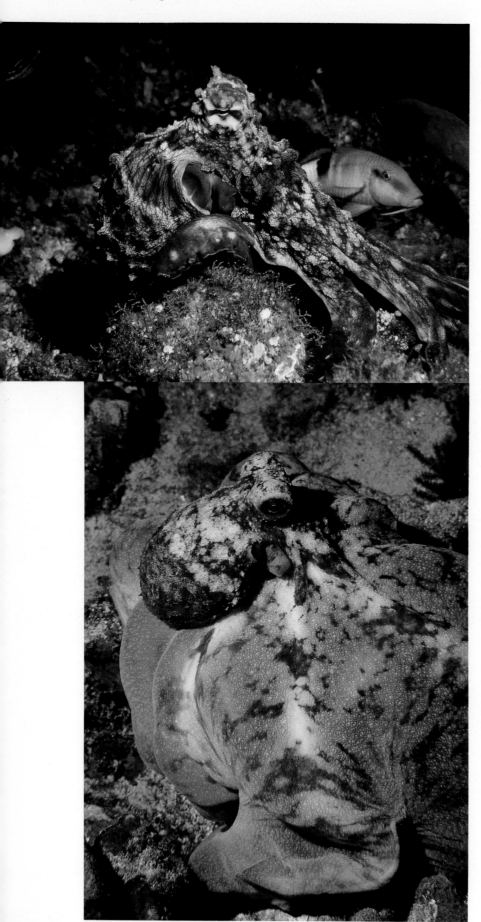

For unknown reasons, octopus display brilliant iridescence at night (Bonaire, Caribbean).

Disguise can also mean losing oneself like this goby in the relative immensity of a red sponge (Grand Cayman, Caribbean).   ▶

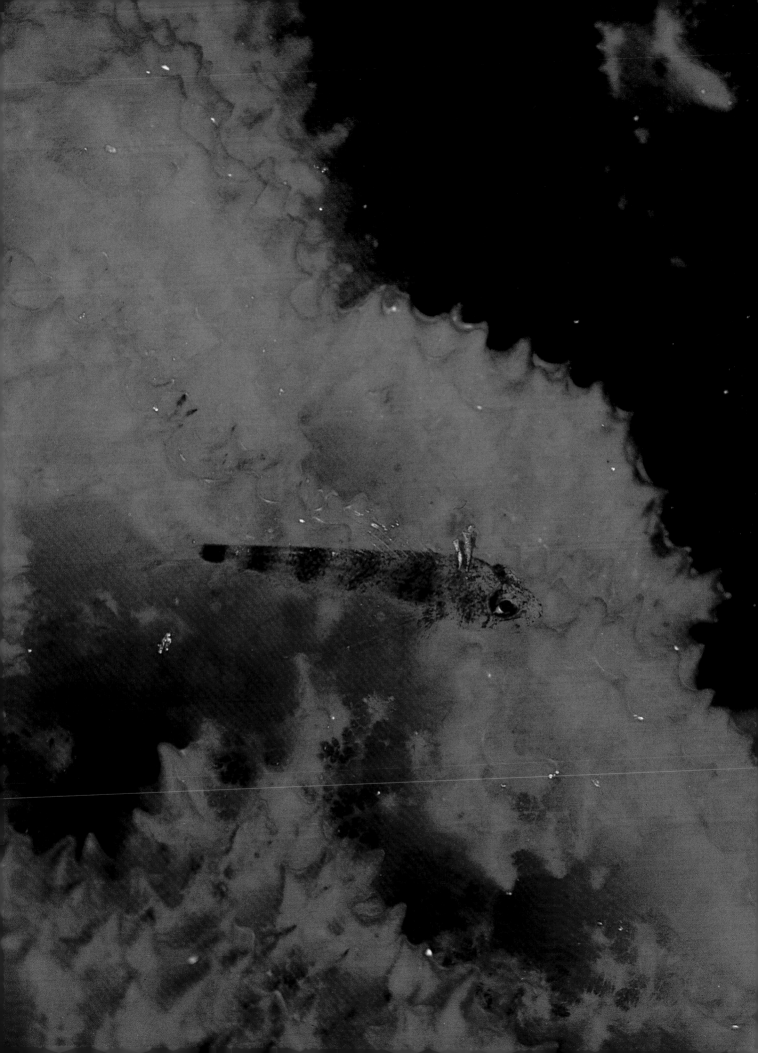

9

Defense

Throughout this volume we have seen a variety of predation techniques. Now and then I have paused to point out that some of these techniques are useful for *defense* as well. After all, for a predator species to survive it must have defense against other predators who would attack it.

Take the octopus of our Masters of Disguise section. Not only can it use its fluid body and kaleidoscopic coloration to reach its hidden shellfish prey; the very same abilities enable it to escape fish or eels that would attack it.

Indeed, the octopus and other cephalopods even have a special defensive technique to escape when they are surprised by an attacker in open water. Lunge at any cephalopod and you will grab a handful of ink, expelled to create a momentary false target.

I say momentary because many defensive techniques evolved by sea creatures seem intended only to confuse or mislead an attacker for an instant, that crucial margin of headstart which allows the prey to escape. Cephalopod ink is a good example of this misdirection, drawing the attacker to the wrong target.

Another example of misdirection is the false eye-spot near the tail of the four-eyed butterflyfish, *Chaetodon ocellatus*. Scientists feel that this false-eye target leads the attacker to lunge toward the tail. Since the butterflyfish's spines point backward, it cannot be swallowed tail-first.

Defenses are as plentiful and varied as the predation which spawns them.

Consider the family of pufferfish and porcupinefish, whose members can inflate their bodies with water to triple their normal size. By making themselves a much larger mouthful, they may discourage all but the largest predators from even trying to bite them. The porcupinefish and spiny puffers (*Diodon*) enhance this inflation ploy with sharp spikes which become erect as the body puffs. Anyone who has ever been careless enough to be jabbed with these spines knows they would fend off the most determined attacker. After all, it takes a large predator indeed to attack a football-sized target covered with two-inch spines.

Inflating is a defense of last resort, for with it the pufferfish gives up all maneuverability; it much prefers to simply duck into a coral crevice until the danger is past. Still, when cornered, handled, or bitten it will inflate to its full girth within a few seconds. Even a predator which has managed to grasp the pufferfish may be driven off as spines rise before its eyes.

In chapter 8 I pointed out that in many reef-dwellers predation and defense techniques are often hard to separate. In the case of the pufferfish, however, their special talent appears to be exclusively defensive, with no offensive component whatever. These fish eat molluscs, crushing their prey's hard shells in their powerful beaks. The ability to inflate, or their powerful spines, would appear to have no offensive role. Another case of a defensive technique with no apparent offensive component is certainly that of the flying fish (*Cypselurus*). Hovering beneath the surface in open blue water, these small, succulent fish are outlined against the bright sky for any of the blue-water fish we mentioned in chapter 6. Barracuda, jacks, tuna, houndfish, and a long list of other candidates soar up from darker waters to devour these flying fish.

The flying fish, however, can in an instant launch itself up above the surface of the water, spread its wings, and "take off," propelling itself with several powerful strokes of its still-immersed tail. Given just an instant's warning, the flying fish will disappear from the predator's view and not alight for a hundred yards or more.

Other defensive techniques may or may not have an incidental offensive role. Consider the many animals which bury themselves in sand. All but the very largest wrasses, for example, bury themselves completely in the sand by night. Stingrays will also immerse themselves in sand, with only their eyes protruding. Razorfish (*Hemipteronotus*) take sand-burial a step further. These small wrasses can actually dive into the sand and move about under it, out of sight of potential predators.

Another graceful fish has put its own individual stamp on the use of sand-burrows. The sand tilefish (*Malacanthus plumieri*) builds an extensive system of burrows with multiple entrances, or failing that, builds multiple burrows. This foot-long slender white fish builds large mounds of coral rubble over its burrows by carrying pieces of rubble in its teeth the way a dog might carry a bone.

When approached from a distance the tilefish will hover, its tail fins streaming behind it like a flag in a stiff breeze. As the intruder moves closer, the tilefish edges to a point above its burrow. A bit closer, and the fish dives under the pile of rubble and is lost to view.

The jawfish (*Opisthognathus*) is another burrow-dweller. It dances in a tail-down position above its vertical burrow, entering either tail-first or (when chased) headfirst. This dainty creature is constantly seen shoring up the walls of its hole by pressing bits of shell and coral in place with its mouth.

In each of these particular instances, sand-burial is purely a defensive adaptation.

Yet other creatures such as the lizardfish (*Synodus*) and the stargazer (*Dactyloscopus*) also bury themselves in sand, and their purpose is purely predatory, to hide themselves from potential prey. Either will rocket up out of the sand to attack small passing fish or crustaceans.

Defense is astonishingly varied. Even motionless corals and anemones defend themselves by means of their poisonous stinging darts, the nematocysts. In chapter 1 I pointed out that the powerful neurotoxins injected by these nematocysts would paralyze a good-sized fish almost instantly.

The nematocyst is, of course, non-selective. Touch its cell (known as the cnidocyte) and it fires; it neither knows nor cares whether you are predator or prey. If it is powerful enough it becomes as good a defensive weapon as an offensive one.

Scale and power are important, however. Coral polyps have nematocysts, yet they are for preying upon plankton. Many fish and invertebrates may browse upon the corals with impunity.

The same cannot be said for the corals' relatives, the hydroids and anemones. These are substantial animals with potent nematocysts, and the coral browsers give them a wide berth.

Only the anemones' exotic symbionts, the shrimp, clownfish, and certain other damselfish, escape doom from the tentacles. Yet there is a predator which does ingest the nematocysts fearlessly. Certain nudibranchs not only ingest the powerful stinging cells, but suppress their firing so these chemical grenades may be swallowed whole. Even more amazing, the unfired cells are then carried to the nudibranch's own external gills, where they become a crucial defensive weapon for the nudibranch itself.

The variations on defense are legion. From the color wizardry of the peacock flounder to the obliterative countershading of the cruising manta ray, the sea's creatures are ever aware that their final predator may be somewhere about at any moment.

Defense can take any physical form whatever. In an earlier chapter we envisioned the hunting eels and sea snakes, threading through the labyrinthine coral reef in search of prey.

In these long, slender predators, defense takes the

same form as offense. The coral fortresses through which they flow so effortlessly are their shields as well as their hunting grounds.

I must confess that this is more true of the various types of eels than it is of sea snakes. Even the large morays are armed only with their sharp teeth, though a six-foot long moray is, from the human perspective, armed awesomely. Still, when most divers see moray eels, what they see is the eel's head peering out of a sheltering coral crevice while its body remains completely entombed in sheltering limestone.

Sea snakes, on the other hand, show absolutely no fear of anything. These slender air-breathers seem to be protected like the Old West's gunfighters by an aura, a reputation. They will swim anywhere, at any time, and nothing in the sea approaches them. Such a defense is a requirement for the snakes, for they must travel to the surface to breathe every hour or so. A lifetime of these trips could be short indeed if predators were inclined to take a crack at you each trip.

So the evolutionary word is out on these lethally armed snakes. At the very first glimpse of the snakes, sharks, tuna, jacks, and other predators simply go on about their business. I've never even seen a predator veer in to eye a snake. In fact, the snakes will swim right into a pack of feeding sharks to grab a morsel, showing not the slightest hesitation. Obviously their reputation is so well established that they are as close to invulnerable as any creature in the sea.

Some fish have morphological features uniquely suited to their defense. The fused-scale, boxlike bodies of the trunkfish (*Lactophrys*) and boxfish (*Ostraeion*) can frustrate all but the most powerful jaws. Even more intriguing is the erectable dorsal spine of the triggerfish. When approached, the triggerfish will wriggle into a coral niche, then raise its trigger and wedge itself into the coral. You may get a grip on its tail, but you cannot get it out short of dismemberment.

In all of these cases, each animal relies on its own burrow, its own coloration, its own morphology, to defend itself.

Other animals develop defensive strategies, based on togetherness, grouping together in large numbers. Near the shelter of the reef, this strategy is best exemplified by such schooling fish as the anchovy (*Anchoa*), the silversides (*Allanetta*), various goatfishes, surgeonfishes, and even the stinging catfish (*Plotosus anguillaris*). Schooling has several purposes, some of which are clearly offensive (you may on occasion see schools of surgeonfish, stinging catfish, or parrotfish swarming over the coral reef, feeding like a swarm of locusts). But despite these occasional offensive manifestations which parenthetically all take place on or close to the reef mass, schooling is essentially a defensive tactic.

One need only see a massive school of anchovy or surgeonfish moving in synchronization to realize how effective a defense schooling can be. I mentioned earlier the school as a mimic of a large single creature; seen from a distance, the school may look as large as a ray or whale. Still, that is a long shot defense against skilled and experienced predators.

The real value of the school is in confusing the attacker's eye in close quarters. You can watch this confusion when a school of anchovy is being patrolled, harassed by swift jacks. The jacks sweep back and forth, trying to isolate a target. Yet the school of fish is so confusing as it ebbs and flows that the attackers make pass after pass and still come away unfed. Indeed, the jacks can swim right through the school and a silver tunnel forms instantaneously around their path. It is both artful and eerily beautiful.

The ability of the school to confuse is mesmerizing; the school flows with every slight move of the predator. You may wave your hand and the school will flow like quicksilver right around it. This is instinctive but extraordinarily skillful behavior, and having observed it closely many times I can attest to its uncanny effectiveness.

While the general purpose of this volume is to examine techniques of predation, it is important to realize that defense is merely the flip-side of predation; move and countermove are constant in the great chess game of the sea. A swimming fish may dodge one predator, only to swim into the mouth of another. That safe-looking, "empty" sand bottom may be packed with hidden predators ready to strike.

Even some of the potent predators of other chapters must adapt some defensive strategy. The crocodilefish *Suggrundus*, for example, resembles sandy rubble. Even the eyes of this amazing fish are covered with a lacy filigree which effectively camouflages them.

The defense may be to resemble something else, the way the frogfish *Antennarius* resembles a sponge, or the way an octopus resembles a coral head, or a flounder resembles sand. One interesting variation on resembling something else is found in the small filefish, *Monacanthus*. The body of this small fish is covered with mottled colors and complex frills which make it seem like a leaf drifting amid the waving arms of gorgonian sea whips.

Allying oneself with a more formidable symbiont is effective most of the time, but the mortality rate among Pacific clownfishes argues that this ploy is certainly not foolproof.

The more one reviews the panoply of offensive and defensive stragems, the more one leans toward the conclusion that the game is incontrovertibly stacked in favor of the offense.

Nature need only provide high birthrates and numerous progeny to make this vast game work. Rather than a delicate balance between life and death (which could be hard to maintain in local catastrophes), the system provides plentiful prey for predators and accepts high losses as natural; indeed, high losses are simply the nutrients of the entire system.

In those few observable instances in which prey populations have diminished (sea lions and sea otters along the West Coast of North America, for example), corresponding reductions in predators have occurred immediately. When prey populations recovered, predator populations did as well. There is at the moment a burgeoning population of seals and sea lions along the North American West Coast due to protective laws; there is a consequent explosion in the population of great white sharks.

This phenomenon is seen on coral reefs as well. Local catastrophes such as volcanic eruptions or heavy rainfall can eradicate entire coral reef areas; the snuffing out of coral life takes out entire populations of browsing fish, crabs, shrimp, and the food chain above them. Yet within a year the "reseeding" of the devastated reef may be seen taking place by the settling out of pelagic larvae. To the human eye, the restocking of a reef in this manner may seem an agonizingly slow process. Nature, however, sees time in millennia, and the process measured in geologic time is really quite rapid.

As the growth of the corals and algae rebuild the base of the food chain, predator populations move back in as well. The sea's predation system is precisely that — a system. It can be disturbed or even locally snuffed out, but the planet wide system is geared to restoring itself after natural disasters. After all, the planet has suffered many cataclysms; without adequate recovery mechanisms there might be no life at all on once-devastated regions of Earth.

Yet with all the powerful drive of predation, overall populations flourish in all seas. This is the ultimate defense, the defense of the system. The system tolerates the loss of individuals in staggering numbers, yet preserves itself.

Science is discovering many instances in nature where a mother deprived of her natural brood will produce another. What a perfect response this is, how simple and yet how appropriate. If the first brood survives in acceptable numbers, the mother's biological system is not triggered so readily — a second brood would merely cause overpopulation. The system demands only that enough progeny be produced to offset predation or other calamity.

Looking at predation from the prey's point of view lets us really appreciate its marvelous structure. We humans are fortunate to be so large that when we snorkel or dive on a tropical coral reef we are larger than all of the local predators. Even we fit a roomy niche in the predation system. This is fortunate, for in this unnatural environment most divers are clumsy indeed when compared to fish, and would certainly be at a disadvantage if they did have to avoid predators.

Whether the human visitor is a predator or not, careful observation of the reef-dwellers in the vicinity will reveal all the natural responses to the presence of a predator. As we move about, some fish swim away, others enter coral burrows, and at the very least all plan their escape — just in case. The more often you dive, the more defense responses you will see. Then, at a certain point in your development you will become so skilled that the reef-dwellers will not be frightened of you. That is when all the mysteries of predation will start to become visible for you.

A small shrimp, Periclemenes
Yucatanicus, *walks fearlessly
amid the poison-armed tentacles of an anemone (Bonaire,
Caribbean).*

◄

Some molluscs such as
Murex scorpio *stand out on
the reef (Maldive Islands).*

The triggerfish Cantherhines
pullus *enters its coral crevice
and wedges itself in by rais-
ing its trigger (Bonaire,
Caribbean).*

When a hermit crab uses a discarded shell for its protection it wants least of all to be noticed (Fiji).

Many reef fish such as fairy basslets hover in confusing profusion to deter predators (Philippines).

Some large pufferfish (Arothron) are too big to fit in crevices, so they nestle in against the reef and eye us warily (Maldive Islands).

*The saber-toothed blenny
(Aspidontus taeniatus) enters
its coral shelter tail-first so
that it can keep an eye on
possible predators (Maldive
Islands).*

The fringed filefish (Mona-canthus ciliatus) *hovers amid the protective arms of gorgonian corals (Bonaire, Caribbean).*

A small goby finds shelter in the color and complexity of a large soft coral (Coral Sea, Australia).

The boxfish, Ostracion tuber-
culatus, *also has evolved its
boxlike body from fused
scales (Coral Sea, Australia).*

The trunkfish, Lactophrys tri-
queter, *has a shell-like body
composed of fused body-
scales (Curaçao, Caribbean).*

*Shaped like small leaves,
these small fish (Aeoliscus)
don't resemble other fish
either in body shape or
locomotion (Papua New
Guinea).*

Not only is this pufferfish able to inflate, it is also disruptively colored for camouflage (Galápagos).

This sand tilefish (Malacanthus plumieri) emerges from the burrow it carefully builds out of coral rubble.

The spiny pufferfish (Dic-
tolychthys punctulatus) is an
unappetizing meal indeed
when it inflates and thus
erects its spines (Fiji).

A cluster of sabellid worms can withdraw into their protective tubes at a hint of nearby motion (Belize, Caribbean). ▶

Woe to the swift fish that tries to bite this serpulid worm and misses. The worm leaves a sharp spine when it withdraws into its tube (Grand Cayman, Caribbean).

The serpulid worm defends its brilliant plume by withdrawing abruptly into its limestone tube (Grand Cayman, Caribbean).

10

Ambush! Sudden Death on the Reef

Of all the predation methods in the sea, ambush is the most sudden and — to our human eyes — the most impressive. It is practiced by a remarkable variety of animals ranging from thousand-pound groupers to the minuscule arrow blenny.

Ambush relies on the element of surprise, and this surprise rests on the inability of the prey to detect the predator until too late.

There are many variations on this theme. The predator may be a massive grouper, hovering in murky water, its vast gray bulk blending into its surroundings. The jewfish, *Epinephelus itajara*, may reach a weight of 1,000 pounds, and presents an awesome appearance as it watches you. Some legends in the South Pacific even credit these giants for the disappearance of native spearmen. Other stories from the oil rigs of the Gulf of Mexico have had commercial divers engulfed up to their waists in the mouths of these creatures.

These jewfish and their myriad smaller cousins often use the structure of the reef as their cover, hovering beneath an overhang or ledge. Countless times I have looked into the reef-shadow to see a pair of motionless but fever-bright eyes staring back at me. The largest groupers can open their huge mouths so quickly that it creates a suction that literally "inhales" small prey from two feet away. The grouper remains in its coral shadow, sucks in its victim, and never even betrays its position.

If, on the other hand, the prey is a bit further away, the grouper will erupt forward with one sweeping stroke of its broad tail. It can cover a distance of two or three feet like a bullet; that plus its inhalation technique assures the instant death of its momentarily careless prey.

One fascinating and merciful side effect of the carnage of ambush is the role of shock. The reef is, when closely observed, a merciless domain. An enormous variety of creatures devour other animals as a matter of daily routine. Nature has provided shock as a merciful buffer; whether the prey has been stung by an anemone or bitten by a grouper, it seems to suffer massive, sudden trauma. This serves two roles — it eases the suffering of the prey, and also protects the predators from injury.

This quick and basically merciful killing can be very important. Sea snakes, for example, have fragile jaws. They compensate for this with powerful venom that quickly overwhelms the sense of small fish, enabling the snake to swallow the fish without injuring itself.

I've watched many ambushers erupt from hiding and attack their prey in one massive charge. At the first bite the prey shudders and is still. There is seldom a struggle to escape. Granted, it is usually a very one-sided contest, with a large animal eating a relatively small one. Still the observations are consistent across a broad range of predators and prey.

While some predators are very colorful to our human eye, their colors usually serve to render them less visible to other creatures in their natural habitat. It is important to realize that in the photographs we take we use bright lights to restore full-spectrum color to our subjects. In their natural monochromatic blue world many of these predators can be nearly invisible against their backgrounds.

Groupers, as I mentioned, can be dull gray to black to blend into their surroundings. Some are colored in shades of red; red looks black in the blue-shadowed world of the reef. Only when we illuminate the fish with our lights does the red emerge; place it in reef shadow and it is as effective as jet black.

Moreover, many species of groupers (the largest and most varied family of ambushers) can change color at will. They may darken when in shadow, or fade to near white when out on open sandy bottom. Other species will develop disruptive patterns of spots or blotches, which are hard to detect against a bottom of coral rubble.

Groupers range from thousand-pound behemoths to the tiny hamlets, and their colors vary from jet black to canary yellow or fire-engine red. They have made an extraordinarily successful career out of mastering the element of surprise.

As a different approach, another group of ambush predators have learned to be unseen even when in plain view. This group, which includes frogfish, scorpionfish, lizardfish, and others, does not hide in crevices or under ledges; they are usually found resting on the reef, in plain view but completely motionless. Most of them betray no movement whatever until the prey is within reach; then in one explosive motion they engulf their prey. Their immobility has rendered them effectively invisible.

There are some fascinating variations on this theme. The frogfish, *Antennarius*, for example, blends this technique of lethal immobility with two other specialized adaptations for ambush.

One adaptation is their color. Frogfishes occur in bright red, canary yellow, orange, brown, and jet

black, and some have spotted patterns. Interestingly, the color scheme of each motionless frogfish precisely matches that of nearby motionless sponges. You can look at one sponge after another — and suddenly one has grim, steady eyes.

A second, even more revolutionary weapon of ambush is the frogfish's lure. This is a small appendage on its nose shaped rather like a fishing pole. Under muscular control the frogfish can raise or even wiggle its lure while remaining otherwise motionless. Any curious small fish which investigates the moving lure is engulfed when the frogfish suddenly opens its capacious maw.

These frogfish exemplify the basic elements of ambush — the comparative invisibility of the predator, the element of misdirection as embodied in the lure, and the overwhelming, lethal suddenness of the attack.

There are other adaptations of these elements, each resulting in an effective predator. Lizardfishes (*Synodus*) and stargazers (*Dactyloscopus*) are capable not only of immobility, but of practicing that stillness while buried up to their eyes in sand. These foot-long, tapered speed merchants may be betrayed only by their eyes until their prey flutters into range. Then, in a swirl of sand, the lizardfish arrows out to overwhelm its chosen meal.

Some of the reef-dwellers we've seen in earlier chapters are also masters of the "fast food" ambush. Among these are the moray eels, with their uncanny ability to insinuate themselves through the honeycomb of reef; the voracious mantis shrimp *Squilla* in its burrow of sand; and the silvery jack, which swoops out of the open blue water like a fighter plane. Each of these we have treated in another chapter based on some defining habitat or characteristic. Yet each also practices the technique of ambush to secure its food.

While somehow we are mentally prepared for a grouper or moray eel to lie in ambush, some other ambushers are totally unexpected. One example is the delicate tartan hawkfish, *Oxycirrhitus*. This long-nosed, delicate fish has a bold tartan plaid body coloration, and is usually spotted propped up on its pectoral fins on a gorgonian fan. The complex backdrop of the gorgonian is a perfect color match for the hawkfish as it awaits its prey.

Because this fish is so small we may be deceived into considering it somehow gentle. Yet the cirrhitids are considered a link between the sea basses and the scorpaenids, both champion ambushers. What really occurs in a human observer is a mental process which equates bright body color and diminutive size with harmless behavior. Nothing could be less valid. You wouldn't think that of a terrestrial spider, for example. You would think of it as a spider first, and as a small and pretty creature a distant second. Yet in reef-dwellers we are lulled somehow into judgments based on a vague prejudice that those pretty little things can't be killers.

Don't bet on it.

Consider for a moment the small wrasse *Epibulus insidiator,* also known as the slingjaw. I've followed these industrious fish around the reef, and succumbed to their air of the busy office worker bustling about corporate headquarters.

There is a unique morphologic feature in *Epibulus.* Its jaw is fully one-third of its body length, and extensible. Now imagine the small prey just ahead of this bustling slingjaw. It may think it has just enough of a lead to reach the security of the coral; until the slingjaw fires its jaws forward and snatches the hapless target. This awesome weapon is hard to believe until you see it in action.

There is another family of predators which human observers tend to misjudge. These are the trumpetfishes, or *Aulostomidae,* which are relatives of the pipefishes and seahorses and which reach a body length of up to two feet. With their long, tubular bodies and miniature horse faces, the trumpetfishes sometimes impress divers as harmless "horses" grazing at pasture.

Totally wrong. These slender predators can hover amid the elongate arms of gorgonian sea whip, then suddenly dart out to overwhelm a passing fish. From the prey's viewpoint, the tiny body-section it sees belies the size of the approaching trumpetfish until it is too late.

These clever predators have an even better trick; they will swim along nuzzled against the back of a harmless parrotfish, almost like a rider hunched over a horse's withers. Then in a flash they will arrow down to engulf some small fish that was lulled by the parrotfish's comforting bulk.

Another impressive predator is the arrow blenny, *Lucayablennius zingaro.* This fish is but two inches long, and indeed reminds one of an arrow. It hovers near the coral reef with its small tail curled tightly to one side. The first time I saw this, I thought the tail might be damaged, but I soon learned by error. The tiny arrow spotted its prey, a small shrimp, and gently maneuvered itself into position with its pectoral fins.

When it was "aimed" properly, it straightened its tail with a snap, firing forward like its namesake. The shrimp never had a chance.

In recounting these few variations on the classic ambush, I have not tried to be exhaustive, merely to portray the reef as it is: a dangerous place for the unwary.

Most small marine species are extraordinarily fecund; they lay hundreds or thousands of eggs. Relatively few of these progeny survive to breed a succeeding generation, for these young are not only the generations to come — they are a principal food supply for the reef carnivores.

An important caveat for human observers is to leave their terrestrial preconceptions behind when they enter the reef habitat. This small horse-faced fish may not (indeed is almost surely not) a harmless grazer. That motionless, small, ugly blob of a fish has evolved that unimpressive mien precisely to blend into its background. Among the ambushers the supreme compliment is to be overlooked — until it is too late.

And watch those shadowed ledges. The groupers hover motionless, their bright eyes fixed on the sunlit water just before them, tails ready to boom them forward for a meal. Patience, uncanny patience is the watchword of these obsessive animals. I was recently on a dusk dive on an Australian Reef; dawn and dusk are the so-called crepuscular periods when predation is at its height.

I was cautiously approaching a small grouper when it darted off to one end of a small coral tunnel. The tunnel entrance was too small for the grouper's head. Like a frantic dog trying to go down a hole, it then darted to the other end of the tunnel, which was also too small. Back and forth, back and forth it went. The animal in the small tunnel must have been terrified during this all-out attack, even though it was encased in rock. The fury of the assault was hard to believe. Like a dog, the grouper tried 20 or more iterations of front-door, back-door, then abruptly gave up with what certainly appeared to be a disgusted look on its face.

Parenthetically, our great fortune as humans visiting the coral reef is our comparatively huge size. Most reef creatures need a substantial size advantage over their prey, simply because most prey is swallowed whole. That means a predator would have to be eight or nine feet in length before it considered us potential prey. Add to that the incredible racket we make in the comparatively quiet reef world, and you see why we are left alone.

Indeed, on a reef in its natural state we would cause all the reef creatures to be extremely cautious. We are an unknown quantity in terms of speed and technique, but our size alone makes every creature on the reef consider itself a potential target.

If we humans were, say, merely a foot tall, entering the reef world as it exists would be fraught with peril. Immediately we would be morsel-sized for most of the groupers, barracuda, moray eels, and other ambushers. In such a world, indeed, most water sports would be nonexistent, for no one would be foolish enough to enter the sea.

Ironically, sensationalist stories have convinced many potential divers that the ocean is indeed extremely hazardous. The truth is, however, that with the single exception of the larger sharks, man himself is the monster-figure on the reef. All of the ambushers, no matter how awesome in their own milieu, tend to shy away at the approach of an unknown diver.

This is not only because of our daunting size, by the way. We are also so noticeable that all small eyes are drawn to us; ambushers, craving anonymity to lie in wait for their prey, tend to be exposed whenever they are near us. For this reason, they will often go to ground and watch us, resuming their stealthy hunt only after our departure.

In one of his short stories, H. G. Wells evoked a stunning image: "In the valley of the blind, the one-eyed man is king." In this world dominated by the hunt, it should be no surprise that stealth and speed are king. The reef-dweller that loses its sight or speed, or fails to recognize certain ambush sites for what they are, is soon a meal.

Is all this cruel? No — no more than our harvesting grapes for wine or wheat for bread. It is merely the way this society functions. A measure of the success of this society is that it flourished for millennia before humankind's emergence. Indeed, many predators, from crinoids to sharks, have remained unchanged for hundreds of millions of years.

Since it is not at all clear that we humans will be that successful ourselves, we might well consider how this reef system has remained so successful and stable. One reason is that it has adjusted its birth rates over the millennia to provide sufficient food for its population. Most reef creatures produce prodigious numbers of progeny. The portion of the food supply that is composed of offspring tracks closely with the overall (and hence the predator) population.

Second, the reef system is built for cleanliness. Below the active predators are a host of scavengers living off the debris of the hunt. On the reef, nothing is wasted.

All in all, this predation-based society is a great practical success; it should be measured on that success without the imposition of an artificial and inappropriate morality.

This rock hind (Epinephelus adcensionis) *is actually in the process of swallowing a parrotfish (Bonaire, Caribbean).*

This red anglerfish
Antennarius multiocellatus
looks black when seen in the
shadows. It lures its prey by
wriggling a filament above its
mouth (Curacao, Caribbean).

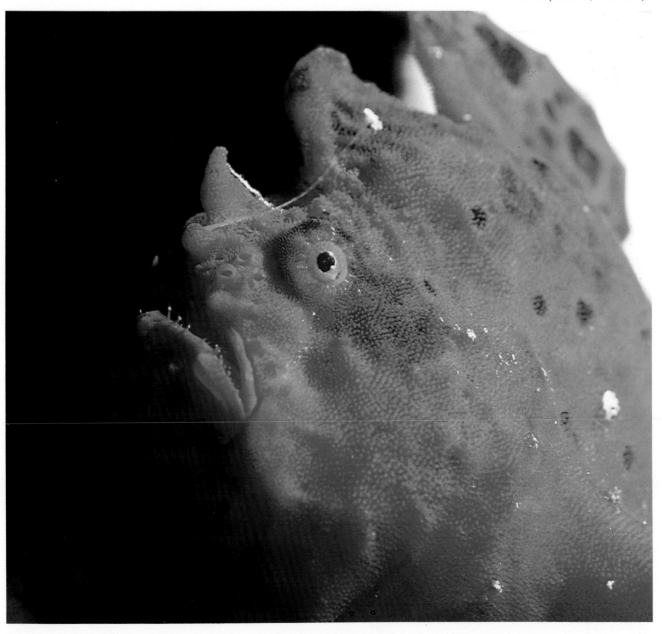

*The deadly stonefish
Synancea horrida lies mo-
tionless on the bottom,
waiting to open its mouth
suddenly to engulf its prey
(Papua, New Guinea).*

The slingjaw (Epibulus insid-ator) has a nasty surprise awaiting any prey that thinks it has a good head start (Palau).

This golden beauty is a color phase of the Caribbean coney Cephalopholis fulva, whose attack is a savage miniature of the larger groupers (Belize, Caribbean).

Like an arrow in a bow, the lizardfish (Synodus) is ready to rocket forward from its coral resting-place to devour unwary fish (Fiji).

The finery of this moray is irrelevant to its nocturnal hunt; perhaps it is for intra-species recognition.   ▶

The extraordinary tartan hawkfish (oxycirrhites typus) perches on a gorgonian (Galápagos).

One of the most bizarrely ornate fish of the reef is the rare scorpionfish Rhinopius aphanes *(Papua New Guinea).*

Small but swift and savage, the coney Cephalopholis fulva is a fixture on Caribbean reefs.

The scorpionfish Scorpaena *sits motionless in shadow, where its red color renders it nearly invisible. When it strikes, its movement is instantaneous (New Zealand).*

This giant Sea bass finds a comfortable resting-place under a rocky shelter (Great Barrier Reef, Australia). ▶

The scorpionfish is a master of not being noticed until its prey has wandered too close (Maldives).

Groupers have an amazing ability to alter their body coloration to blend into their background. This is Epinephelus fuscoguttatus *seen against an irregular background (Fiji).*

The Indo-Pacific spotted grouper, Cephalopholis miniatus, *is an intense red, a color which looks black in the monochromatic blue world of the reef (Maldive Islands).*

The striped sweetlip (Plectorhynchus cuvieri) looks too innocent to be a predator, but it is (Australia).

11

Children of the Night

Night on the coral reef is a time of profound change. Even as the sun goes down the change begins; to the human observer the transition betrays visible signs while the sky is still light.

The first indications of the onset of night may be increased activity among groupers and schooling fish in the darkness beyond the reef dropoff. This period before sundown and darkness is often the most exciting of the day in the numbers of larger predators one may see.

At the same time, smaller diurnal reef fish stay much closer to the reef surface than they do in earlier daylight. No one wants to be silhouetted against the lighter surface water, for this is the dreaded crepuscular period. At dawn and dusk, predators hidden by darkness rocket upward to take prey made visible against the sky.

Literally hundreds of times I've sat on a dive boat during these hunting periods and watched the water churn as a school of baitfish is attacked from below. Sometimes the predator — jacks, tuna, or mackerel — actually arc right up out of the water in the fury of their crepuscular assault. The silvery anchovy leap out of the water in graceful waves of undiluted panic, only to fall back into danger. Wave after wave of silvery prey roll across the mirrored surface, with nothing to save them but their sheer numbers.

Underwater near the reef mass, many of the wrasses, angelfishes, butterflyfishes, parrotfishes, and other daytime reef-dwellers are suddenly conspicuous by their absence. Other diurnal fish such as pufferfish hover in protective crevices looking out anxiously at the darkening waters.

In human terms, there is a whiff of the citizenry behind closed doors during a crime wave. Of course, it is not a crime wave, merely the predators' period of most intense activity. Still, from the viewpoint of these anxious citizens peering out from hiding, the crime wave analogy may not be too far off.

Soon, looking upward, we see that the last light has left the sky. With our hand lights, we probe the reef and find it totally altered from what we witnessed in daylight. For one thing, some familiar faces are missing altogether. The wrasses are gone, as are the parrotfishes, butterflyfishes, and angelfish.

As we prowl the darkened reef, we'll find parrotfishes and butterflyfishes nestled in coral crevices that offer them almost complete protection. The angelfishes, so fearless by day, are almost never seen by night, apparently because they penetrate deep into crevices in the reef structure.

The wrasses are totally out of sight as well, for their nocturnal strategy is burial in the sand. They are joined by some pufferfishes, lizardfishes, flounders, and others.

Snappers and grunts neither bury in sand nor hide in crevices. They gather in small groups and hover motionless in small coral valleys between gorgonians and stony coral heads. When we shine our lights on them, even from a distance, they stir restlessly and move off through the coral valleys in search of undisturbed sleep and the anonymity of endless darkness.

Ocean triggerfishes, normally seen by day soaring in open blue water off reef walls, nestle into depressions on the irregular face of the vertical wall and pancake themselves into place; surely no free-swimming predator could detect their presence. In a similar ploy, hawksbill and green turtles may find themselves coral depressions to nestle into, or they may hover unseen within the sheltering arms of a large gorgonian sea whip. By daylight, most of these stratagems would seem ridiculous, for the animal is still in plain view. One comes away with the strong impression that nocturnal hunters succeed by pure chance; move about the reef poking your head into depressions, and you'll inevitably find some prey to eat.

One point of great interest is the nocturnal color change many diurnal fish undergo. This phenomenon is hard to miss, as you will see by examining several of the photographs that accompany this chapter. This color change phenomenon is noteworthy first because there is no apparent reason for it, and also because it is shared by a substantial number of species.

In some cases the color change is modest. In the threadfin butterflyfish (*Chaetodon auriga*), for example, there is merely some darkening in its rear dorsal area. Yet other butterflyfishes, such as *Chaetodon ocellatus*, develop bold patterns which totally transform their daytime appearance. Several questions are inevitable: If these fish are resting in darkened crevices, unseen, why change color? In some cases such as the porgy (*Calamus*) and the glasseye (*Priacanthus cruentatus*), the shift is from solid body colors to blotches which may be disruptive to the predator's eye. But in the dark?

One is forced to speculate that the vision of nocturnal hunters is so acute that these patterns may be useful in what appears to us complete darkness.

Alternatively, these color patterns could be a throwback to an earlier epoch when they were useful, but have now perhaps become anachronistic. Or there may be some completely different reasons which have thus far eluded us.

We are haunted, though, by the sheer fact that all of these color alterations take place in what surely seems total darkness. "Total" darkness may of course be misleading. I have dived in crystal-clear tropic waters when the moon was full, and been able to navigate the reef without a handlight. I was not able to make out these hidden or color-changed reef-dwellers, but then I am far from what an effective nocturnal predator should be.

Be that as it may, when we probe with our bright lights, some of the color changes we witness are little short of astonishing. One need only to stumble upon a sleeping goatfish (*Pseudopeneus maculatus*) to see vividly what I mean.

By day this is basically a white fish with three black squares on its flank. What reason could it have for shifting to complete red body tones with accents of iridescent blue — especially since it usually rests on open white sand, where the red color would seem to make it contrast with the sand rather than blend into it?

Consider further that this fish has a second, daytime use for this nocturnal color scheme. When a school of goatfishes approaches a cleaning station, one individual will often shift to nocturnal colors to stand out among its companions. The cleaners invariably offer first service to the color-shifted individual. How did this signal evolve?

This mystery of unseen color is a broader question, by the way, than merely the color-shifting of a few nocturnal species. There is a larger mystery in the fabulously bright colors of many deep-water gorgonians, soft corals, stony corals, and even fish, whose brilliance occurs in the perpetual night of the deep reef 200 feet or more below the surface. On reefs all over the world I've flashed my lights on creatures whose raiment defied description — yet without the full-spectrum light of a strobe or handlight they are hardly visible at all.

Obviously, this phenomenon is one which deserves thorough study. For now, the uses of color in marine species raises more questions than answers and the conundrum of night color shifts are but a piece of that larger puzzle.

For surely, if there were some general benefit in

changing color at night, *all* of the creatures should do it. Further, if prey are going to change color, why don't the predators change, too? Why not have red or iridescent sharks or eels? In this Renfieldesque world of eerie darkness, the children of the night know how to sing, but we do not understand their music. . . .

If so many of the diurnal fish take night-refuge of some form, is the reef left empty? Not at all: An entirely new cast of characters rises to take their place.

One group we've already discussed at length in chapter 3. The walkers as a group are almost all nocturnal feeders. There are certainly both offensive and defensive aspects to this nocturnal feeding. Many of the walkers are coral-predators, and the best feeding opportunities are at night when the polyps are extended for their own feeding. Not only the coral polyps, but most of the small crustaceans, molluscs, and open-water wrigglers are abroad at night, offering a plentiful and varied food supply to the predatory walkers. We will often discover lobster, crabs, and shrimp moving out over the reef surface with the impunity of clanking armored knights, their claws and mandibles constantly at work.

This suggests not only that these predators find their richest feeding at night, but also that the predators who hunt them are either inactive or less effective in darkness. Considering how thoroughly hidden these crabs, lobsters, and shrimp may be by day, I suspect that one or more of their major fears is large wrasse or triggerfish which are absent by night.

There are two major groups of reef fish which become active in darkness. One is that of the squirrelfishes, soldierfishes, and bigeyes. Another is the variegated family of cardinalfishes (*Apogon*).

The squirrelfishes, soldierfishes, and bigeyes share certain unmistakable characteristics. They are red in body color (at least by day), and their eyes are oversized compared to other fish. These huge eyes are structured to offer greater light-gathering characteristics, and hence enhance night vision. By day, these red fish hover in coral crevices and under ledges, motionless and inactive, though still wary of the intruder. At night they move up into the waters above the brooding dark bulk of the reef, where they pluck plankton and tiny fish from the passing food stream.

One of the night's marvels is the effect observable when you shine a powerful light on a glasseye (*Priacanthus cruentatus*) or its cousin the bigeye (*Priacanthus arenatus*). Part of your lightbeam impinges on the huge eye and is immediately reflected out again. If you are directly perpendicular to the eye's interior reflective surface, you'll get the stabbing beam right back into your own eye — a sure case of the punishment fitting the crime. We never realize what discomfort we inflict upon the nocturnal fish with our hand-beams until we are subjected to it ourselves.

If you or the fish moves slightly to an angle, however, you'll see a brilliant miniature of your beam jetting out into utter darkness.

These large-eyed fish hover close above the reef in darkness, picking tidbits from the food stream with little competition. That's because many of the plankton-browsers of the daylight hours, such as grammistids and fairy basslets, are sound asleep in their rocky beds. Succulent small crustaceans wander the reef feeding, only to be engulfed without warning by the mouth beneath the great, lustrous eyes. Predation, after all, is a game that never ends — there are just different players sitting in to play some hands.

As distinctive as the big-eyed red fish are the *Apogonidae*, the cardinalfishes. This family includes members that are less than an inch long and totally transparent, and others that measure up to six inches long and are boldly striped from nose to tail. All have a distinctive eye-bar, and their eyes are large for their body length.

During daylight hours the cardinalfishes hover beneath coral ledges in the same manner as the soldierfishes and squirrelfishes. When darkness comes they move up into the open darkness above the reef, snatching bits of plankton from the open water.

The utter blackness of a reef at night is an experience difficult to express in words. The ultimate way to experience it is to be alone and have your light fail. If you are on an active reef in New Guinea or Australia's Coral Sea far from any hotels or civilization, you know in that moment that man has not driven away the large predators, and you feel entirely vulnerable. Your imagination peoples the inky vastness about you with unimaginable horrors, and for a moment you feel perhaps like a small cardinalfish or squirrelfish — a very quick meal for something monstrous that this way comes. . . .

Diving in front of a hotel beach on a busy Caribbean island, you may be calmed by your daytime experience of never seeing any fish over a foot long; on a primitive, wild reef there is no such comfort. The darkness becomes endless, oppressive. You populate the sea with creatures far more fearsome than are really there. Every story you've ever heard about large animals is suddenly front and center in your mind, and the cold fingers of uncertainty tap your shoulder.

Is this only our own overheated imagination, or is this our closest approximation of the lives these nocturnal creatures lead?

Personally, I'd guess that marine creatures lead lives of wariness but not of dread. For one thing, dread is an emotional state peculiarly tied to human imagination; most animals (and some men) seem to endure fear only when directly faced with danger.

There is a second reason for my conclusion: These creatures would be literally incapacitated if constantly plagued by dread. From an evolutionary standpoint, an incapacitated species is a species doomed to extinction. This may be a warning to humans as well, when fears of conflict confuse and incapacitate us. For a species to function it must pursue its activities with full faculties, not with part of its attention fixated on the unthinkable.

That said, we humans may enter the nocturnal reef with a new mind-set; it is now merely a reef that is dark, where some of the locals are asleep and others are awake, but where the endless game of predation is still in progress.

Given that freedom from dread, we may open our sense to the intricate and color-splashed tapestry of the nocturnal reef.

I was recently diving a reef in the Coral Sea at night which exemplifies that tapestry. We were diving a large atoll whose rim nearly reaches the surface. That means the only access to the lagoon for tidal waters is through various cracks or passes in the rocky rim.

During this series of night dives we explored a pair of crevices perhaps 15 feet wide. We could feel the warm waters of the lagoon flowing out around us, a rich food stream indeed; as we played our lights on the rocky walls we saw they were literally covered with soft corals, gorgonians, crinoids, and night-blooming corals in a stunning profusion. So complete was the settlement that nowhere could one see the rocky coral wall that supported this life. Here and there moved a red crab, or a bright-eyed shrimp, or a massive nudibranch, or a soldierfish, or a flared lionfish, or a bigeye, or a cardinalfish. Nestled down between corals were sleeping pufferfishes, parrotfishes, and butterflyfishes. The food concentration caused by the current flowing through this crevice caused a profusion of life that seemed limited only by the available substrate. Each creature, after all, had to have someplace to hang on. Some, such as the crinoids, would daintily crawl up on a gorgonian coral formation and achieve unobstructed feeding. Others — small corals, sponges, and hydroids — seemed almost lost in an avalanche of life.

We had been told that one of the superstars of the nocturnal reef frequented these crevices, so I turned off my lights. To my delight, the entire crevice walls blinked with tiny, flashing green lights. That's right — hundreds of the flashlight fishes (*Photoblepharon*) were scattered in small crevices along the walls.

The flashlight fishes are distributed all around the tropical world. Once thought rare, they are now known to be everywhere, though visible only under limited circumstances.

Mounted on the flashlight fishes' cheeks just beneath their eyes are small pouches in which flourish colonies of bioluminescent algae. The brilliant light is an intense blue-green, and is produced by a natural luciferin-luciferase reaction similar to that of the common firefly. The flashlight fishes almost certainly use these pouchlights to feed, illuminating plankton at close range. Scientists speculate that the lights are also useful for finding each other for breeding, and perhaps even for attracting prey or for signaling.

The signaling theory arises from observing the flashlight fishes' ability to "blink" its lights on and off. In *Photoblepharon* and *Anamalops* a shade or lid is drawn up to hide the glowing colony; in the Caribbean species *Kryptophanaron alfredi* the entire pouch is rotated inward into a receptacle. Both mechanisms turn the glowing pouch of algae into signaling-lanterns which the fish may blink at will.

While we do not understand all of the uses of this unique light-show, we find ourselves lost in wonder as it turns the reef wall into a flickering Great White Way, a miniature Broadway in the ocean's darkness.

The sea at night. Vast. Dark. Misunderstood. Populated with our own demons. When we transcend our fears we discover it to be instead a phantasmagoric wonderland of exotic color and wondrous form. It is a domain to be sought out, reveled in, rather than fearfully avoided. For many divers it is a frontier they have not yet crossed.

For those who hesitate, the cliché says it all: *Try it, you'll like it.*

*This parrotfish is wedged,
motionless, into a
noctural reef (Red Sea).*

◀

*Day and night colors of the
spotfin butterflyfish,
Chaetodon ocellicaudus
(Belize, Caribbean).*

The night colors of the goatfish Pseudopeneus mac-ulatus. *In the background a lizardfish lies buried in sand (Virgin Islands).*

Day and night colors of the butterflyfish Chaetodon xanthocephalus; *are the spots to look like two eyes peering from a hole? (Fiji).*

At night on open sand one may encounter anything. Here is a tiny juvenile *scorpionfish* (Scorpaena) (Galápagos).

This is the extraordinary flashlight fish Anamalops, with its cheek-pouches filled with bioluminescent algae (Philippines).

Day and night colors of the
threadfin butterflyfish,
Chaetodon auriga *(Fiji).*

Nocturnal anemones festoon themselves upon a long wire coral to achieve unobstructed access to the food stream (Fiji).

Day and night colors of the glasseye Priacanthus cruentatus (Philippines).

Parrotfish (Scarus) *tend to nestle into natural coral niches (Fiji).*

The lionfish Pterois volitans, hovering motionless against the coral reef, may be mistaken for a harmless crinoid (Philippines).

Night and day colors of the porgy Calamus bajonado *(Grand Cayman, Caribbean).*

◀

The dainty cardinalfish (Apogon) *are nocturnal replacements for daytime feeders such as grammistids and fairy basslets (Philippines).*

The Indo-Pacific squirrelfish, Holocentrus spinifer, is the world's largest (Maldive Islands).

The Caribbean squirrelfish, Holocentrus rufus, *is reclusive by day, rising to feed after dark (Belize, Caribbean).*

Some of the arms of this nocturnal basket starfish (Astroboa nuda) *have already captured prey. This is the spiderweb of the sea, except that the web is itself living (Cozumel, Caribbean).*

The Classic Predators — Sharks

12

In the opening sentences of the introduction to this book, I made two observations: That mankind has conditioned itself through eons to fear sharks, and that sharks for their part constitute the incarnate symbol of the predator.

Do sharks deserve their fearsome reputation? Certainly not all of it. Movies such as *Jaws* which ascribe dark motives of personal vendetta to these animals owe far more to metaphysics, mythology, and ignorance than to science.

That said, it is also a dangerous error to consider all sharks creampuffs as some divers do. Any large predator can be a hazard under the right circumstances, and it is in the peculiar nature of sharks that those circumstances occasionally converge.

It must first be admitted that despite ever-increasing study of sharks and their behavior, there is much about them we simply don't know. That is because you cannot truly comprehend a large, open-ocean predator by watching it in a small pond. The didactic approach betrays itself; the subject taken out of context becomes a different subject. The method of observing can itself directly alter the behavior being observed, which may spawn precisely the wrong conclusions.

Moreover, there are dynamics of the open-water environment itself that arise in certain situations which are impossible to duplicate even in the most elaborate captivity.

One reason is that it is hard to generalize about sharks is that there are so many different kinds of sharks. Small epaulette sharks or Port Jackson sharks are bottom-dwelling scavengers who spend much of their time reclining on sand or rubble; even larger nurse sharks and wobbegong sharks are seldom seen swimming unless we have disturbed them. These scavenger sharks are not the dark figures of our mythology; only novices are so shark-conscious as to fear these placid, even lazy animals.

No, the paradigmatic image of terror is reserved for the requiem sharks (*Carcharhinidae*), the sleek, dark, open-water prowlers of the endless sea. The problem is of course that we fill in the wide gaps in our own knowledge with hearsay, gossip, and pure myth, until we can't tell where the knowledge ends and the gossip begins. Not long ago, for example, some of my friends were collecting stories of shark attacks for a book. After sifting through dozens of accounts, they realized they were hearing about the same few attacks again and again, duplicated and embellished for local consumption.

Why is that? For the same reasons that I get calls and letters from photographic agents asking for any pictures I can provide of "vicious sharks." In other words, please provide photographic materials to fit certain preconceptions. You may detect a process at work here. The prejudices and fears of the editor select illustrations which prove his or her point — the myth thus creates "evidence" to prove the myth.

Sharks are *not* vicious — they are simply effective predators. So effective are they, in fact, that we have come to project our fears upon them with broad strokes of soul-chilling imagination. In a single day I have heard reports of a shark encounter or attack already being enhanced, embellished, elevated. The shark grows larger, darker, and more savage with each retelling.

Does that mean sharks are really pussycats? Definitely not. They are more like big jungle cats, another terribly misunderstood group of predators.

Requiem sharks and other open-ocean varieties spend much of their time prowling extended ranges looking for food. Targets of opportunity are often ill, injured, or merely old — those who have lost an edge of awareness or agility that separates life from sudden death.

Open-ocean sharks, which include blues, mako or mackerel sharks, oceanic white-tips, hammerheads, and others, roam far afield and are seldom or never seen near coral reefs.

Another group, which includes nurse sharks and gray reef sharks, is composed of residents of the reefs with large ranges or territories which they ceaselessly prowl.

Coral reefs do not offer particularly rich feeding for open-ocean sharks. First, most of the reef creatures are small and well-adapted to seeking quick refuge in the maze of the reef mass. In addition, any ill or aging reef-dweller is more likely to be eaten by another nearby reef-dweller before the far-ranging shark arrives on the scene.

If, however, a "trigger" disturbance occurs on the reef, the situation is altered radically. This alteration occurs so rapidly that human observers may be taken completely off-guard. One moment the sharks are patrolling sentinels, the next moment they are a writhing mass of violent attack.

Trigger disturbances may take several forms, which the shark detects by sound, smell, or sight. Given the short range of sharks' vision in sea water, they rely primarily on their other senses to detect and locate feeding opportunities. One key is low-frequency vibrations emanating from struggling fish. Spearfishermen have long known that the sounds their catch makes often brings hungry sharks. This type of sound carries long distances in open water, often exceeding a mile.

One of my own experiences amply illustrates this.

I was feeding a group of sharks as a filming opportunity for diver clients in Australia's Coral Sea. Our bait was fish caught the day before, so we were not making noise to draw sharks; we were merely bringing food to a dozen sharks known to frequent this reef area. While there was intense activity as the sharks fed, any sound broadcast was merely that of a pack of sharks feeding. Even those sounds certainly should have been obscured by the exhalations of a dozen divers.

Nevertheless, as I cleaned up various ropes and wire after the feeding, I was suddenly confronted with a 15-foot hammerhead shark.

It is important to note that large hammerheads are open-water predators, and are not normally found near this type of reef. Yet here it was, its predatory instincts raised to fever-pitch by the sounds it had heard.

How far had it come? I'll never know, of course, but what was impressive was that it came unerringly. From the distance, it tracked the sound; as it drew closer it undoubtedly detected blood and fish-oil in the water; from a hundred feet or so away it utilized its keen vision. Those three senses orchestrated this immense shark over perhaps a mile until it flared to a majestic stop not five feet from me.

It was, to say the least, disconcerting. Fortunately, this behemoth ate some fish-heads I promptly dropped out of my hand. If my luck had been worse, it might have chosen me as its meal.

It did not, however — adding one more incident to a long history which indicates a possible general pattern of behavior: Unless triggered into a frenzy, sharks seem to take only the safe meal. I've done many feedings with packs of gray sharks; invariably these sharks will rocket around, between and over our divers but always make their way to the fish-carcasses which are the bait. In these shark-feedings there is no sense of real danger, more a sense of being with a pack of dogs hungry for a meal. Like dogs, the sharks in these situations seem easily to distinguish between humans and bait. What change of circumstances, then, would make sharks lethal to humans?

Three changes would make the difference, I think: Change of venue, change of shark, and change of trigger.

By change of venue, I specifically mean moving from the reef mass out into open water. For reasons we may only dimly perceive, sharks in open-water situations seem to perpertrate attacks on humans which might not have occurred in a reef environment. I've had this experience myself on three occasions.

On one of our shark feedings, I made the error of re-entering the water after lunch near the site of our morning shark feeding. Our boat was farther from the reef, and when I stepped off the dive platform a current swept me away from the boat. The sharks, which in the morning were like a pack of noisy but controllable dogs, altered their behavior immediately. As soon as they perceived me separately from the boat they began to race about the bottom at 60 feet, then formed into a funnel-cloud shape and rose to attack. I bashed sharks on the nose with my two big cameras, back-pedaling furiously until I reached the reef wall; at this point the sharks abruptly broke off the attack.

On another occasion, for the rapacious demands of a TV film, I and some other misguided souls jumped into the middle of a pack of feeding blue sharks. We were 12 miles off San Diego, in open ocean, and by bad luck the water visibility was less than 20 feet. There we were, in pea soup laced with snapping blue sharks, when suddenly a mako shark joined the fray.

After the mako attacked one of our cameras, our safety diver was forced to kill it with a bang-stick. Without our safety diver riding shotgun, there seems little question that mako shark would have hit one of us. As it was, the turmoil raised the level of intensity so high that we had to get out of the water. Who knew what would show up next?

Finally, there is my favorite, the oceanic white-tip shark, *Carcharhinus longimanus.* This splendid, sleek prowler cruised up as I languished aboard a dive boat which had temporarily broken down.

For an hour, miles from shore, three of us swam with this ocean-roamer in crystal-clear, calm Red Sea water. Its pattern was patient and irresistably persistent. It made long, lazy circles about us, always watching. The circles gradually became smaller, and after a while I had to bang the shark on the nose with my strobe light after each picture. Without question, if we had not been able to exit the water, we would eventually not have survived the encounter.

In each of these cases the change of venue, change of shark, or change of trigger heightened the risk of the situation.

The change of venue from reef to open water can be seen to increase the danger because one is moving from an environment where the shark usually fails to one where it usually succeeds. In the open sea there are no coral crevices to dart into; you must outswim, outmaneuver, or outmuscle this creature which is so perfectly adapted to hunting in this environment.

This is especially true when you change sharks from gray reef to mako or hammerhead or oceanic white-tip. These royal sharks spend their entire lives searching these open waters for prey. They are amazingly proficient both at finding potential prey and at overwhelming any prey they do find.

Finally, we should consider what I've called the trigger, that one element in the situation which turns any shark or group of sharks into attackers.

Prominent among these triggers is noise, particularly loud, sudden, and low-frequency sounds. Humans thrashing about like fish thrashing about, and noises such as an explosion or the impact of a plane crash will bring sharks to an attack point very quickly. Sharks have an extraordinary sound-detection and location system; along their lateral line is a fluid-filled channel lined with sensitive cilia which gives the shark an uncanny ability to home in on sounds being propagated through its water medium.

Even unfamiliar low-frequency sounds such as the backwash of helicopter blades has been known to gather and trigger sharks. This caused some anxious moments in several astronaut recoveries over the years.

I'm convinced, for example, that the fifteen-foot hammerhead I mentioned earlier sought out the sounds of the smaller sharks feeding, perhaps from a mile or more away. Considering how the noise from a dozen divers' exhalations should have masked the feeding sounds, that hammerhead displayed an amazing discrimination in analyzing and tracking the sounds it heard. The powers of detection and location are so acute that we whose senses are practically vestigial cannot even imagine having them.

Another well-known trigger is blood or fish oil in the water. It doesn't take much. I've seen sharks follow unerringly the trail of a wounded, bleeding fish as if it were a white line on a highway. There is no guesswork; sharks are so sensitive to these odors that they can follow them for miles, another ability we can but dimly conceive.

One of my favorite expeditions is that for the great white shark (*Carcharodon carcharias*) in the cold waters off southern Australia. To attract these open-ocean wonders, we set a bucket of whale oil with a slow leak over the side of our boat. To this attraction we add some scrap tuna meat replete with blood and oil. This creates a greasy slick which drifts out for miles in the moving tides.

Sooner or later a wandering great white shark crosses this spoor. Unerringly following this trail, the shark comes to us, whereupon it puts on the most impressive display of predation I've ever witnessed.

Throughout this volume we've seen predation techniques of many kinds; yet this primordial creature makes one forget all the rest.

For one thing, it is huge — 12- to 15-foot great white sharks weighing a ton or more are still adolescents. Secondly, these sharks are so beautifully camouflaged with obliterative countershading that you find it hard to distinguish them even if you know where to look. Thirdly, you never do know where to look, because these large sharks possess an unbelievable instinct for coming from the blind side. Finally, given all those skills for getting close to their prey, the attack of these large sharks is an awesome display of sudden, savage power.

I've watched them take a 30 pound chunk of horse-meat, bones and all, and chew it up as we would a bite of steak. I've also watched them attack bait-cans, plastic floats, the corners of our shark-cages, the propellors of our boats, boarding-ladders, and other delicacies.

What, then, is the difference between the great white shark and the others I've mentioned? I believe it is a difference in degree. All of these predators cruise the open seas following prey. In the case of the great white shark, its principal prey are wary and fast-swimming sea lions and other pinnipeds. To successfully prey upon these alert, swift mammals the white shark must have the advantage of surprise.

One natural advantage for the shark is the cold, usually turbid waters it shares with its prey. Add those impressive adaptations of color, mass, and instinct for the unexpected approach, and you have sketched the ultimate predator in the sea.

Think of creatures such as the large hammerhead or the great white shark, and it is understandable, even inevitable, that Man has developed a primordial fear of these animals. They rule a vast world in which we are thoroughly incompetent; they embody the ultimate penalty for our temerity in roaming sea and sky in our fragile and fallible craft. They know nothing of mercy, doubt, or remorse. They are totally alien to us, creatures from the ancient past who were so perfect they have ruled for a quarter-billion years. Even the dinosaurs looked out from the shore and saw the dark shadows in the sea.

Yet we must not become the victims of our own loquacity. On the reef, divers are in fact extremely safe. They may enjoy seeing sharks merely as the apotheosis of marine predation, in the end the most beautiful of all the sea's creatures.

Enjoy the beauty, forget the fear. Behold the perfect predator.

This large nurse shark Ginglymostoma, was reluctant to leave its comfortable resting-place (Coral Sea, Australia).

This nurse shark Ginglymostoma has had its fin bitten off, and swam away quickly when I disturbed it (Palau).

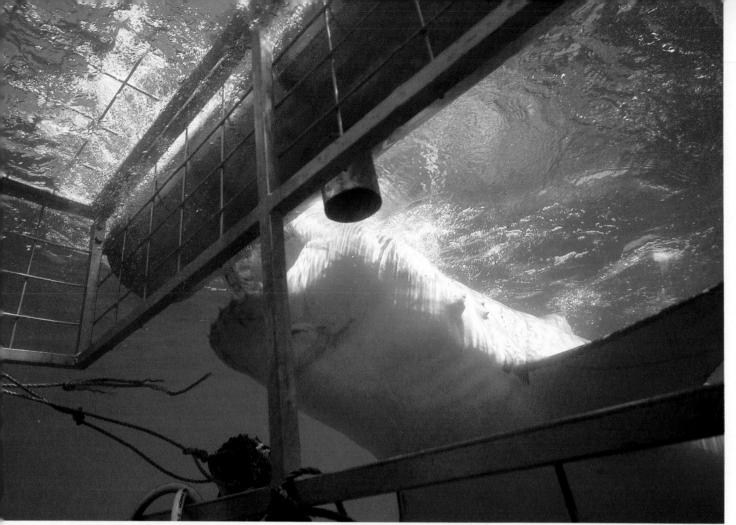

Even more attention-getting is its penchant for trying to take a bite of our cage (South Australia).

Once the great white shark begins its bite its power is awesome (South Australia).

There is no thrill to match watching the great white shark display its ferocious artistry (South Australia).

Lord of the open seas is the magnificent oceanic white-tip shark, Carcharrhinus longimanus *(Red Sea).* ▶

The carpet or wobbegong shark (Orectolobus ornatus) *is elegantly disguised but very lazy (Papua New Guinea).*

Not all sharks look sinister. The Port Jackson shark (Heterodontus quoyi) *sits motionless on sandy bottoms in cooler regions (Galápagos).*

Rising placidly from the sand, a cat shark (Stegostoma fasciatum) *soars out into open water (Coral Sea, Australia).*

Conclusion

Man — The Final Predator?

We cannot consider marine predation without briefly considering the role of man in the sea. Oddly, much of mankind's most serious impact in marine species is not really intentional predation, but merely the by-product of our complex, lethal, and unheeding civilization.

Why is man's assault on marine life so effective, and hence so damaging? I first began to consider this question years ago, when I learned to spearfish with experts from Curaçao. These lean, powerful divers would hold their breath, drift along in a current with guns cocked, and unerringly ambush large groupers. After a while, the few surviving groupers grew wary, correctly identifying these human hunters as dangerous predators.

This is a case in which evolution failed the grouper. Instead of out swimming the relatively inept humans, the groupers used their ancient defense against other fast-swimming marine predators — they went to ground, hiding in coral crevices where sharks could not follow.

Tragically for the groupers, this new species of predator used a weapon (the speargun) against which going to ground was precisely the wrong defense. One by one they died, trusting in their coral fortresses and being ignominiously shot where for millions of years they had found safety.

Over the years I have seen many other of man's assaults on the sea in much the same way — they are innovations in predation for which marine life has not yet evolved a defense.

Powerful tuna boats with motorized winches and miles of net were unknown for hundreds of millions of years as man's prey evolved their defenses. So were miles of floats with baited long lines hanging in open ocean. So were chicken-wire fish traps, trolling-lines, baited hand lines, throw-nets, poison (such as coke-bottles filled with lye), and other strategems.

Man's lively imagination has produced these innovations, but has often done so without fully realizing the havoc these effective methods can wreak.

Fish traps and nets are unselective. They kill whatever they catch. After long use, they may draw the feeding population of prey down to a non-sustaining level. Soon, the fishermen find their catches dwindling as their own success destroys their livelihood.

I understood this when my spearfisher friends wistfully looked out over plundered, empty dive sites and mused, "You know, ten years ago we could get five or ten big groupers on every dive here. . . . " The Law of Unintended Consequences is alive and well in the world's seas.

In the Philippines, boats with hundreds of small boys smash coral reefs with stones to drive the resident fish into nets. Slowly, they are pulverizing miles and miles of reef, then moving on.

Whalers still pursue those endangered monarchs, tuna fleets now use aircraft as spotters. We are gradually pursuing some species to their final haunts; what will we do when they are gone?

Even worse than human predation are the side-effects of civilization itself on the seas. Human wastes have made large areas of sea-floor totally devoid of life. Chemical wastes in particular have become a symbol of civilization gone amok. Finding DDT in Antarctic penguins, or sea birds with thin-shelled eggs that break to the touch, should terrify us because we share our future with them. If we are poisoning the world of these remote creatures can we not see that we are poisoning our own as well?

To my knowledge, no predator species in nature has ever been its own executioner, the victim of its own activities. Man may be the first.

Before the damage passes the limits of repair, man must curtail his destructive activities. It is a relief to note that in some quarters progress has been made. Altering the use of nets in tuna fleets has greatly reduced the unintentional killing of porpoises, who drown when tangled in the nets. Many pesticides have been banned or limited, though we don't know whether this was done soon enough. Rivers that had become poisonously foul have been cleaned, watersheds protected, estuaries preserved. It is a promising start.

Like the old legend of the boy with his fingers in the dike, however, progress does not mean we have won. Victory in one battle makes us realize that other disasters are in progress. The plight of the Chesapeake Bay, for example, may have progressed beyond repair. Those destroyed reefs in the Philippines may take 50 years to repropagate.

We humans have one powerful ally: The regenerative and restorative powers of nature are far more powerful than we ever imagined. Left alone, areas which have been poisoned will be cleansed, reefs which have been devastated will bloom again.

The danger to us all is that man, in his rush for conquest, will persist in his predation until he has overwhelmed the sea's restorative powers. All our human lives, for example, depend upon the phytoplankton in the sea; it is the planet's principal remaining oxygen supply since we have devastated the world's forests. That means simply that an agricultural chemical which attacked the phytoplankton could snuff out all oxygen-breathing life on Earth.

Man the predator would have consumed himself.

It need not go that way, but the bells are tolling. Noisy, preoccupied man had better stop to listen.

Index